CONTENTS

Preface to the Second Edition   *vii*

Preface to the First Edition   *ix*

1   Introduction   *1*

2   The Structure of the Cell Membrane   *12*

The Permeability of Cell Membranes   *22*
Differential Centrifugation   *34*

3   The Cytoplasm   *36*

The Endoplasmic Reticulum   *43*
The Nature of Microsomes   *55*

4   Mitochondria   *65*

The Chemical Nature of Mitochondria   *78*
Metabolic Substances Found in Mitochondria   *78*
Plastids and Chloroplasts   *97*
Lysosomes   *103*
The Metabolism of Carbohydrates   *108*
Fat Metabolism   *129*
Protein Metabolism   *133*
Plant Cells   *137*

5 The Golgi Apparatus   *140*

6 The Nucleus and the Nucleic Acids   *161*
Chromosomes   *172*
Spindle Fibers   *179*
DNA, RNA, and Genes   *182*
Nucleic Acids of the Nucleus   *191*
Synthesis of Protein by Nucleoli   *212*
Nucleocytoplasmic Relationships   *219*

7 Specialized Cells: Gland Cells, Muscle Fibers, and Nerve Fibers   *236*
The Striated Muscle Fiber   *246*
The Nerve Fiber   *270*

8 Conclusion   *284*

Subject Index   *289*

DIVISION OF LABOR IN CELLS

Second edition

GEOFFREY H. BOURNE
Yerkes Regional Primate Research Center
Emory University

ACADEMIC PRESS
New York and London

ACADEMIC PRESS, INC.
111 Fifth Avenue, New York, New York 10003

United Kingdom Edition published by
ACADEMIC PRESS, INC. (LONDON) LTD.
Berkeley Square House, London W1X 6BA

LIBRARY OF CONGRESS CATALOG CARD NUMBER: 74–86362

PRINTED IN THE UNITED STATES OF AMERICA

Preface
to
the
second
edition

SINCE the first edition of this book was published, there have been tremendous strides in the area of cytology. The spectacular breakthroughs in our knowledge of RNA and DNA, and the cracking of the genetic code are but a few examples. Overall, our knowledge of how cells are made, and how they function at the subcellular level, has progressed to the point where it seems feasible to present a new edition of this book.

I have attempted to include in the second edition of "Division of Labor in Cells" at least highlights of the major developments that have taken place in the field of cell biology over the last eight years. In order to keep the book to a manageable size, some portions of the first edition have either been shortened or eliminated.

Some of the author's studies referred to in this edition have been carried out with the assistance of Grant FR-00165 of the National Institutes of Health, NASA Grant No. NGR-11-001-016, and Grant No. HE-04553 of the National Heart Institute.

The author is greatly indebted to various authors for permission to use illustrations from their published work. He is especially indebted to the editor of *Rassegna Medica* for the use of many outstanding illustrations. The sources of other illustrations are indicated in the legends to the corresponding figures.

Preface
to
the
first
edition

DIVISION of labor in cells is a study that provides intriguing problems for both experimental and theoretical scientists. During the last few years new cytological techniques have been developed which have fundamentally altered our approach to research in the field.

Through use of these techniques considerable light has been shed upon the complexities of cell structure and function, but so much remains to be learned that in some instances discoveries serve chiefly to open up new vistas for research and to point out still more challenging unresolved problems.

This volume is concerned with the relationships between structure and the chemical and biochemical composition of cells in general and of several specialized types of cells. It has been planned as a brief synthesis of certain aspects of modern cell biology that will provide some biochemical background for graduate students in biology and anatomy and a morphological basis for graduate students in biochemistry. It is hoped that it will find a niche in the libraries not only of general biologists but also of pathologists and others who are interested in cell biology.

Division of labor is evidenced by the way in which the various parts of the cell perform important but different functions according to the task for which each is structurally and chemically suited. For example, the nucleolus synthesizes a considerable amount of protein and ribonucleic acid; oxidative activities occur in the mitochondria with the consequent production of ATP which plays its vital role in glandular, muscle, nerve, and other cells. A striking factor is the degree to which all parts of the cell cooperate to build up an ordered and controlled metabolic picture. Among the promising areas for further investigations are the mechanism of control of the endocrine system over the cell, the relationship of vitamins and other nutritive factors to the physiology of the cell, and the role of the cytoplasmic (endoplasmic) reticulum.

The author was encouraged to prepare this small book because of the many requests which he received for reprints of a paper on "Division of Labor in Cells" which he presented at a conference on the Chemical Organization of Cells sponsored in August, 1958 by the Pathology Study Section of the National Institutes of Health and later published in the *Journal of Laboratory Investigation.*

The subject of cell biology is one that is advancing with considerable speed and there have been significant advances since the book was written, the reader will therefore not find references to "messenger RNA" as such, its function and significance are referred to but not under that name.

The work referred to in the text which was carried out in the author's laboratory was performed with the aid of the following grants: Muscular Dystrophy Associations of America Inc., No. B-1914 and No. B-2038 from the National Institute of Neurological Diseases and Blindness, No. A-3090 and No. A-2050 from the National Institute of Arthritis and Metabolic Diseases.

The author is greatly indebted to Dr. Henry Hoberman, Dr. Keith Porter, and Dr. Maurice Sandler for reading the manuscript and for many helpful comments.

DIVISION OF LABOR IN CELLS

Second edition

ONE
Introduction

FAR away in the distant past, when the primeval seas bubbled and steamed, when ultraviolet light of great intensity streamed onto the surface of the earth, molecules of polypeptides, of long-chain fatty acids, steroids of various sorts, and purines and pyrimidines ultimately to form part of the nucleic acids, went through the seething pangs of synthesis. Both purines and pyrimidines have recently been found in the interior of meteorites—thus they are not unique to this world and indicate that nucleic acid synthesis may be occurring elsewhere in space.

These compounds, floating free in a warm soup, performed no organized activity. They reacted indiscriminately with each other, blindly, with no form or purpose.

How these complex molecules normally formed only by living organisms came to be synthesized originally without life is a real detective story—parts of which are just now beginning to be filled in. The conception that living things were necessary for the formation of organic compounds was a dogma only a little over one hundred and fifty years ago. One of the best-known organic compounds at that time was urea, found to be excreted by nearly all life forms, but in 1828 the chemist Wöhler duplicated one of life's processes by synthesizing urea in a test tube. Since then of course myriads of organic compounds have been made in the laboratory, including some highly complex compounds.

The fundamental materials necessary for life are protein on the one hand and nucleic acids on the other. The synthesis of these compounds in the laboratory, if it is to be a feasible method for their formation in nature, must be by a method that could in fact have occurred at some time in the earth's history. There is no doubt that in the early stages of earth's development the atmosphere differed greatly in composition from the present atmosphere. Methane, ammonia, and water vapour were present. There was also a high temperature and frequent high-tension electrical discharges through this mixture in the form of lightning. Dr. Stanley L. Miller who, in 1953, was working in the laboratory of Nobel Prize winner Dr. Harold Urey in the University of California, assumed that this is what actually happened and exposed mixtures of methane, ammonia, and water to lengthy periods of electrical discharge. As a result he produced mixtures of amino acids, sugars, and vegetable acids. Since then many laboratories throughout the world have confirmed and extended his experiments.

Further consideration of the composition of the primeval atmosphere has resulted in the conclusion that in addition to methane and ammonia it almost certainly contained hydrogen, hydrogen sulfide, and hydrogen cyanide. It should also be considered that in addition to electrical discharges this atmosphere may have also been subjected to intensive radiation of various forms. As a result of using atmosphere of this type various workers have now produced amino acids, nucleic acid bases, and sugars. It is a long step, however, from producing amino acids to producing proteins. Theorists in this area have included such distinguished personages as J. B. S. Haldane and A. I. Oparin who have conceived of life having developed in the primeval seas. This concept has been criticized recently by various authors who have pointed out that the amount of these essential compounds formed from gases is very small and their dilution in the seas would make it impossible for the polymerization necessary for the production of proteins and nucleic acids to occur. They believe life is much more likely to have arisen in small pools that were continuously drying out. In this way concentrations of molecules could be formed that could initiate polymerization. From these pools that probably were situated near the sea, the primitive groups of active molecules that could be described at this stage as life forms would have been washed into the sea where they could develop further (see Fig. 1).

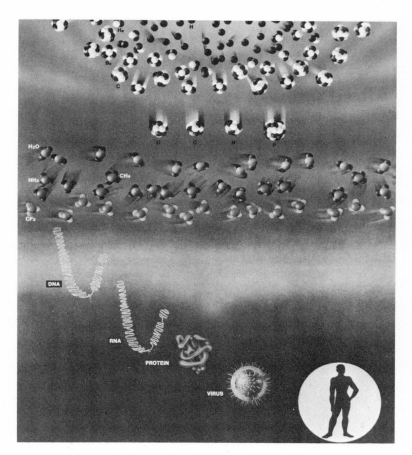

FIG. 1 Oparin and Beadle have put forward hypotheses concerning the basic steps along the path which may have led to the formation of living matter over a period of billions of years. This figure summarizes these hypotheses. In the beginning, two free hydrogen atoms (H) combined to form a helium atom (He). During this process, a neutron (shown as a white ball) is attached to form part of the new atom. Two helium atoms then combined to form a beryllium atom (Be). The beryllium atom attracted a helium nucleus and was transformed into carbon (C). The other elements were formed by similar reactions, among them oxygen, nitrogen, and fluorine. These elements and others react with carbon giving rise at first to very simple compounds, e.g., Methane (CH_4), water (H_2O), carbon trifluoride (CF_3), ammonia (NH_3), etc. More complex substances were then formed until the formation of the first DNA and with that the cycle of living matter began. (From *Rassegna Med.* **41,** No. 3, 1964.)

Matthews and Maser [*Nature,* **215,** 1230 (1968)] have pointed out the key role played by hydrogen cyanide in the primeval atmosphere. They found in their laboratory that the condensation products of hydrogen cyanide and ammonia plus water were capable of forming up to 14 different amino acids, the essential building blocks of protein. They also found something even more important, that many of these amino acids were already joined together in chains (polymerized)—the first step in protein formation. Some of the amino acids necessary for building up proteins contain sulfur and aromatic side chains and, of course, none of the amino acids produced by Matthews and Maser contained either sulfur or aromatic side chains. However, it is fairly certain that if the gas mixture used had contained hydrogen sulfide and acetylene, some of these types of amino acids would have been reproduced as well. In fact, one of the sulfur-containing amino acids—methionine—was synthesized last year (1968) from an aqueous solution of ammonium thiocyanate irradiated with ultraviolet light. Ammonium cyanate (a solid material) is itself formed by passing an electrical discharge through a gaseous mixture of ammonia, water, methane, and hydrogen sulfide. It is possible that fairly large chains of amino acids (polypeptides) could have been formed directly from the gas mixtures. In fact Dr. Philip Abelson, director of the Geophysical Laboratory of the Carnegie Institution of Washington believes that it would be easy for simple proteins (which would be simply fairly long polypeptide chains) to be formed directly. He quotes as an example the enzyme protein "ferredoxin," which has only 55 amino acids. Since the complex chemistry of the cell depends upon the action of proteins in the form of enzymes, it is obviously of great importance that some of the simpler enzymes might have been formed directly from gaseous mixtures. Dr. Sydney Fox of the Institute for Molecular Evolution in Miami has suggested that the heat of volcanoes or streams of lava reacting with gas mixtures may well have provided the energy that resulted in amino acids being formed. He has produced many of these compounds in the laboratory using high temperatures.

In the living organism, as we will see later, proteins are not produced by the same drastic processes that gave birth to them sometime prior to 3 billion years ago. It is of interest that fossil evidence has indicated that biochemically complex organisms were in fact in existence that long ago. In living organisms proteins are formed more gently but much more subtly by building up the amino acid polymers on templates called nucleic acids. Nucleic acids are made up of organic bases, sugar molecules, and phosphate groups; and presumably the components of the first nucleic acids were formed from components that were born in the same violent manner as the first amino acids. The organic bases that form the nucleic acids are adenine, guanine, cytosine, and uracil, and recently Ferris, Orgel, and Sonchez [*J. Molec. Biol.* **33,** 693 (1968)] have been able to produce all of them by subjecting a gaseous mixture of nitrogen and methane and other gas mixtures to an electrical discharge.

However, before these bases can link together to form a nucleic acid they need to have phosphate attached to them—a process known as phosphorylation. Most inorganic phosphates will not react with such bases to transfer their phosphate but, under the influence of heat, such simple phosphates will combine to form polyphosphates, which will transfer phosphates to the bases described above.

One other molecule is needed to link up the phosphorylated bases to form nucleic acids. This is a sugar known as ribose and its related form deoxyribose. By means of chemical reactions using formaldehyde, ribose has been made in the laboratory by chemical methods compatible with those that might have occurred originally in nature. From ribose the production of deoxyribose presents no problems. We are well on the path, therefore, to demonstrating that both protein and nucleic acids could have been produced in the primeval waters of the earth. It is possible that the synthesis of these products could only have been possible at one stage in the earth's evolution—a time of great heat, an atmosphere with just the right combination of gases and water vapor when great sheets of lighting swept constantly through the clouds they formed. Maybe Venus is in a stage today similar to that through which the earth passed when the first molecules necessary for life were forged.

Before protein there could be no life—as Engels once remarked, "Life is the mode of existence of albuminous bodies."

At some stage following this or possibly coincidental with it, chlorophyll was developed. This is a pigment that can use directly the energy of light to synthesize chemical substances; in present-day plants starches and sugars are formed in this way. Whether chlorophyll development occurred before or after the first living organism we cannot say for certain, although its synthesis was probably subsequent to their development.

The formation of proteins, nucleic acids, etc., was not in itself enough to produce living organisms—as J. D. Bernal has said, there has to be a "passage" from a mere living area of metabolizing material without specific limitations into a closed organism which separates one part of the continuum from another, the living from the nonliving. In other words groups of molecules responding to electrical and surface forming forces became oriented in a way that produced a membrane.

Presumably these membranes that were forming on the surface of the water were at first simply flat structures, but "one day" through interreaction of the various molecules a membrane formed a little vesicle. Many other vesicles became formed, and here we had for the first time the potentialities for the development of living structures, for the membrane cut off the contents of the bag from the environment. What could have been the composition of such membranes? Perhaps the amino acids were formed into proteins at the time that this happened. Proteins, in general, have a remarkable property of spreading in a very thin film upon the surface of water—even when they are themselves in solution they are capable of forming these thin films. Compounds such as steroids and fats, i.e., compounds that have a preponderance of hydrocarbon groupings, which makes them only slightly soluble in water, also have the ability to spread on surfaces and to have a particular orientation so that their polar groups will, in general, be in contact with water or with other polar groups, and those parts that have hydrocarbon groups will be out of contact with the water. Furthermore, lipoprotein films which combine both these two types of compounds, can even be formed artificially, for instance cholesterol and the protein gliadin have been used to produce such a film. These films have considerable elasticity and a good deal of strength.

Within the little vesicles we have mentioned, the reorganization of the molecules went on, and eventually a vesicle formed within the parent vesicle, which formed a structure similar to the nucleus. In this a substantial percentage of the nucleic acids became concentrated. Combinations between proteins and nucleic acids formed a basis for inheritance of some of the characters of the cell in a way we do not yet understand fully. By a slow and gradual process of evolution, a complicated structure we now know to be a cell was built up and from it developed the multicellular organisms, animal and plant. At the stage when the multicellular organisms were evolving, it is not even certain that the cell was completely developed in the way we know it at the moment, it probably had a much simpler character than present-day cells.

The bags or vesicles contained proteins, fats, phospholipids, carbohydrates, minerals, enzymes, vitamins, and water. Now if one normally shook such components up in a little bag not much would happen, although no doubt there would be some linkup of some of the molecules to form structures. For example, if collagen is disaggregated with citrate and then is left to stand after the citrate has been removed, many of these groups of molecules reform into collagen units or even collagen fibers. Something of this sort could occur in the type of vesicle we have described if the right molecules got together. On the other hand, in the beginning very little activity characteristic of the living cell would develop. In the presence of their substrates, for instance, many of the enzymes present would have some action, but for some of the enzymes the pH would be wrong and in such a mixture there would probably be only one uniform pH and the temperature would not be optimal. There would be no mechanism for removal of reaction products from the site of their formation, and the reactions would be rapidly saturated with them and would grind to a stop.

The secret of the complex activity of the living cell lies in the isolation or partial isolation or even the timed isolation of its various activities. This is achieved by the presence of various membranes. The nuclear membrane guards the activity of the nucleus, and the Golgi apparatus and the mitochondria have membranes that permit them to carry out activities in partial but not complete seclusion. Figure 2 shows the structure of typical cells, the lower illustration shows a single cell under the high power of the microscope. Figure 3 demonstrates in diagram form the elements of the cell as seen under the electron microscope. Figure 4 shows the last stages of cell division.

It is of interest that natural, fresh water frequently contains films and these are probably derived, according to Goldacre, from decomposed biological material. They appear to contain protein or lipoprotein and in many ways possibly resemble the type of film just described from which cells were probably originally formed. It is of further interest that films such as these, upon collapsing, can spontaneously form vesicles or cylinders, e.g., the wind may cause a collapse of the film. In the case of water passing under

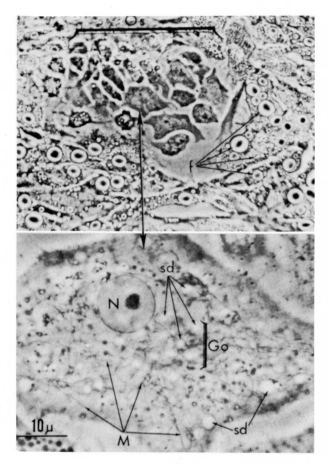

FIG. 2 The living cells under the optical microscope. Migrating cells from two-week-old embryo chick leg bones after three weeks of cultivation under strips of cellophane. The top illustration shows a nest of osteoblasts (Os) surrounded by a dense stroma of large fat droplet (f) containing fibroblasts. The lower photograph shows one osteoblast under high-power magnification. Specific parts of this cell are the nucleus (N), the Golgi complex (Go), secretory droplets (sd) emanating from the Golgi complex, and long filamentous mitochondria (M). (Preparation and photograph by G. G. Rose.)

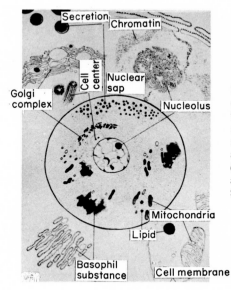

Golgi complex

Cell center

Secretion

Chromatin

Nuclear sap

Nucleolus

Mitochondria

Lipid

Basophil substance

Cell membrane

FIG. 3 What the electron microscope shows. The fine structure of the Golgi complex, the nucleolus, the basophilic substance, and the mitochondria are shown diagrammatically. (From De Robertis, Nowinski, and Saez, "General Cytology," Saunders, 1960.)

a floating barrier, for example, when a stream disappears underground, the surface film is carried toward the point of entry, and according to Goldacre, may be concentrated. Compression of this film causes a wrinkling, and if the wrinkles become sufficiently pinched the walls finally touch underneath the wrinkles and complete cylinders may be detached that separate from the rest of the film. These cylinders themselves can form vesicles that would resemble the primeval vesicles previously mentioned; some may contain air and some the underlying water. Sometimes instead of cylinders, fibers may be produced. It is of interest that the properties of these collapsed lipoprotein films (which can actually be produced in the laboratory) have, according to Goldacre, many similarities to those of the membranes of living cells.

As will be shown in the next chapter, membranes appear to be made up of repeating units of lipoprotein; when such repeating units become disaggregated they tend to recombine spontaneously to form vesicles. It is of interest that the phospholipid itself will not form vesicles—the presence of protein is necessary. Amino acids and proteins existed in the soupy pools of the earth for a long time, perhaps before membranes formed; on the other hand, it is possible that as the former substances were produced many

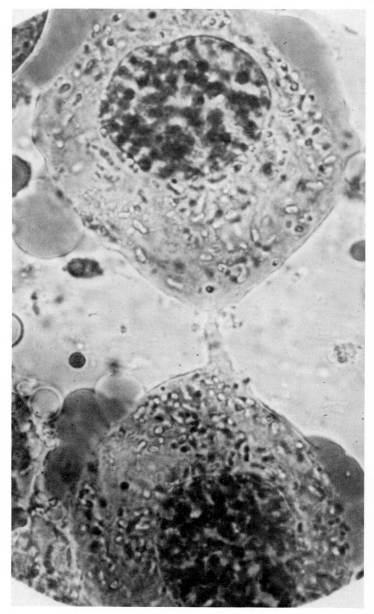

FIG. 4 Sarcoma cells in tissue culture in last stages of mitosis photographed by ultraviolet light. (From R. J. Ludford, Pathology of Cells, *in* "Cytology and Cell Physiology," 2nd ed., Oxford Univ. Press, 1951.)

of them became organized into surface films that became membranes and subsequently divided up into vesicles. As mentioned earlier, artificial membranes can easily be formed, and mixtures of fatty substances, alkalis, proteins, and inorganic salts are very suitable for this purpose. Hereira has even formed in this way what he describes as artificial amoebae that were persuaded to produce types of ameboid movement. But, as Goldacre points out, such movements are uncontrolled, and what is so important in cells is the fact that there is a very considerable degree of control by the cell over itself. Nevertheless, the studies of Hereira help us perhaps to understand how objects comparable to cells, in other words, isolated vesicles containing macromolecules could have been produced in the first place.

Goldacre himself points out that the essential step in the evolution of the cell was the formation around a chemical system of what he describes as a relatively impermeable envelope in the beginning. But, what was presumably a relatively simple structure at first must have undergone a considerable amount of further development, and membranes of present-day cells are relatively complicated structures that take quite an active part in regulating the compounds and ions that go in and out of the cell. For this purpose they seem to have absorbed, adsorbed, or incorporated enzymes in some way onto or into their surfaces, and these enzymes play a part in the process of active transport. Goldacre's concluding comment, however, was that "It appears that the cell uses and may be found in the future to be using mechanisms that are not very different from processes which are already available in the surface films of nature."

TWO
The structure of the cell membrane

THE introductory chapter has suggested that both protein and lipoid substances* may be concerned in the structure of cell membranes.

Let us trace briefly the origin of our present knowledge of the structure of cell membranes. Overton, in 1895, was originally responsible for the idea that lipids were important in cell-membrane structure. He made this suggestion because of the relative ease with which lipid-soluble substances penetrated the cells. At that time there was no clear idea as to what types of lipid molecules were present or as to how they were arranged. In 1917, Langmuir and other workers made deductions about the arrangement of molecules in molecular films spread on water surfaces, and these have been described as being of fundamental importance to the modern concepts of the cell membrane structure. Langmuir suggested (as mentioned in the introduction) that lipids at an interface of air and water arrange themselves in a monolayer with the polar ends of the molecules directed toward the water interface and the nonpolar ends at the air interface. In 1925, Gorter and Grendell found that the total amount of lipid they could extract from

---

* *Lipoid* is a generic term used to include ether- and alcohol-soluble compounds such as phospholipids, fats, fatty acids, cholesterol, and other steroids. Lipid is often used for phospholipid.

a red-cell ghost† when spread in a monolayer occupied about twice the area of the membranes of the red cells. They suggested, therefore, that the lipid was arranged in the membrane as a bimolecular leaflet in which the hydrophilic polar groups were at the inner and outer surfaces and the hydrophobic carbon chains were directed toward each other in the interior of the membrane. This was a simple and attractive model. It is difficult to imagine, however, how such a simple structure could account for all the various functions of the cell membrane.

Dr. J. D. Robertson has described the next stages of the development of our knowledge of the cell membrane. Briefly these are as follows: Harvey and Shapiro in the 1930's and also Cole during the same period studied the surface tension of starfish eggs and came to the conclusion that, at the range of pH's which apparently occurred inside cells, most pure lipoid substances gave surface tension values of the order of 5 dyn/cm^2 or more. But intracellular oil drops were found to have a surface tension of only 0.2 dyn/cm^2. The surface tension of whole starfish eggs as estimated by the force required to compress them is even lower than this. Thus, if only a purely lipid/water interface is concerned, these low values for the living cell cannot be explained. Danielli and Harvey demonstrated that there was some substance on the cell membrane that was responsible for such low values. This substance was a surface active agent of a cytoplasmic nature and appeared to be protein. Apparently in natural membranes all lipid polar substances were covered with at least a monolayer of protein. Danielli and Davson eventually proposed that cell membranes in general had the same sort of structure and that this consisted of one or more bimolecular leaflets of lipid with each polar surface having on it a monolayer of protein. (See Figs. 5–7.) Recent work has shown that these protein layers are probably not more than 20 Å thick, which means that they consist of not more than the thickness of a few polypeptide chains.

In 1936, Schmitt, using polarization microscopy, concluded that at least in red-cell membranes the lipid molecules that were present had their carbon chains radially oriented. Thus we can picture the cell membrane as being composed of two layers of lipid molecules radially arranged (at right angles to the surface of the

† Red-cell ghosts are obtained when red cells are hemolyzed and so lose their contents, and the empty membranes (the ghosts) are centrifuged out of solution and analyzed.

Exterior

Lipoid

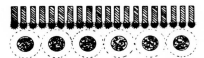

Interior

FIG. 5 The structure of the cell membrane. This photograph illustrates the pauci-molecular theory of Davson and Danielli. The center layer is lipid. The peripheral lipid molecules on both sides are oriented at right angles to the surface. Superimposed on them is a monolayer of protein (spheres). (From Davson and Danielli, "The Permeability of Natural Membranes," Cambridge Univ. Press, 1943.)

membrane) and a monolayer of protein applied to both the inner and outer surfaces of the membrane, with the long axes of the molecules lying parallel to the surface.

It had been suggested in the past that the cell membrane contained either pure lipid or pure protein, with pores about the size of a large molecule, or that it was in the form of a mosaic

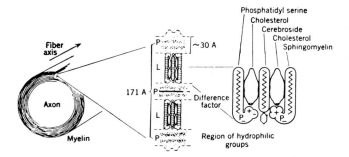

FIG. 6 Molecular structure of the myelin sheath as suggested by Schmitt, Bear, and Palmer. (From *J. Cell & Comp. Physiol.* **18**, 31, 1941.)

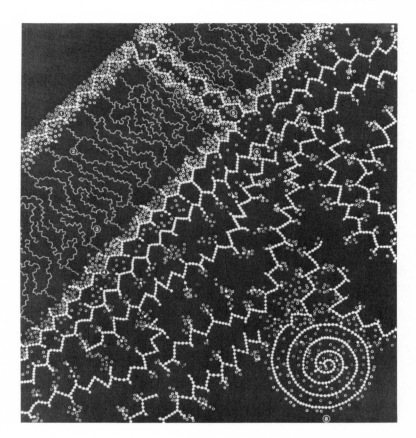

FIG. 7 Diagram of the cell membrane. The cell membrane consists of a double layer of lipid molecules covered by two layers of proteins. The lipid molecules are parallel to each other and perpendicular to the membrane. The nonpolar groups (hydrophobes) of lipid molecules face each other, while the protein chains forming the outer and inner part of the cell membrane are adsorbed on the polar groups (hydrophils). Since each lipid molecule is about 30 Å long and the thickness of each protein layer is 10 Å the thickness of the cell membrane is about 80 Å. 1. Outer protein layer, 2. outer lipid layer, 3. inner lipid layer, 4. inner protein layer, 5. membrane pore, 6, filamentous protein chain, 7. bridges joining the protein chains, 8. globular protein molecule. (From *Rassegna Med.* **45,** No. 3, 1968.)

containing either pure lipid or pure protein in various areas. However, this seems unlikely now except in some special cases. It is of interest that a common structure for the cell membrane exists in cells as diverse as erythrocytes, axons of nerve cells, muscle fibers, leucocytes of the blood, yeast cells, algal cells, the cells of higher plants, and the ova of the echinoderms, e.g., sea urchins and starfish. It seems very likely, however, that there are variations to some extent between the membranes of the various cells; nevertheless it is highly probable that a general structural pattern does exist for all of them. A typical example of variation was demonstrated by Mudd and Mudd, who showed that erythrocytes are preferentially wetted by oil, and leucocytes by water; this indicates that molecules at the surface of the membranes differ.

Studies with the electron microscope (see Figs. 8 and 9) demonstrated that plasma membranes had a thickness of the order of about 80 Å (10,000 Å $= 1$ μ) so that they were about ⅟₂₅ μ thick. Dr. Keith Porter has pointed out that the first electron-microscope photographs demonstrated the cell membrane as a solid structure, but that, as the technique improved and better resolution was obtained, the membrane was seen to be more complex.

At first the apparently solid membrane appeared to be composed of two lines, each measuring 25 Å across and separated by a space of 25–30 Å, making a total thickness of 75–80 Å. It became generally assumed that the two lines represented the inner and outer protein layers and the space in the center the lipoid layer. This structure was shown to be characteristic of the membranes of the cell, of the nucleus, the endoplasmic reticulum, of the Golgi apparatus and the mitochondria and was described by J. D. Robertson as a "unit membrane." In 1962, however, Sjöstrand and Elfvin published an article in which they showed that the structure of the plasma membrane could be changed by the fixation and embedding procedures used. The original preparations of the plasma membrane that were examined under the electron microscope had been fixed in osmium tetroxide (osmic acid) and embedded in methacrylate. Various improvements in embedding and fixation, for example fixing in potassium permanganate and embedding in araldite, showed the original solid structure of the plasma membrane to be symmetrical and to be composed of three units—two 25-Å-thick lines and the 25- to 30-Å space between them. Then it was shown that the application of uranyl acetate or lead staining or the

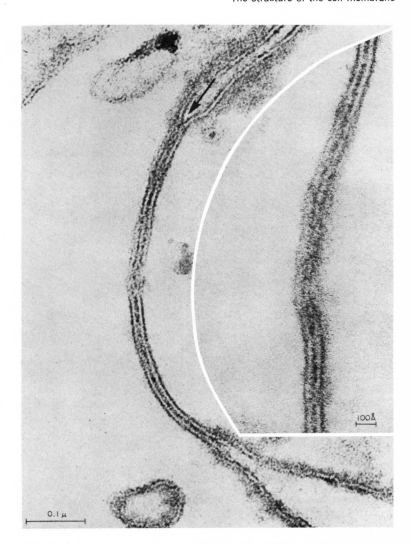

FIG. 8 Plasma membranes isolated from the liver. (From E. L. Benedetti and P. Emmelot, Membranes Isolated from the Liver, *in* "The Membranes," Academic Press, 1968.)

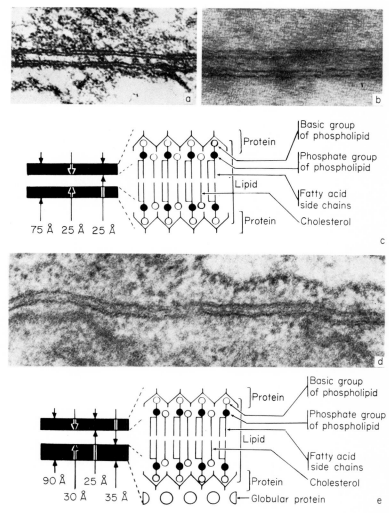

FIG. 9 (a) Opposed plasma membranes of adjacent cells in skin of *Littorina*.
(b) Opposed plasma membrane of adjacent cells from smooth muscle of
mouse intestine. Both these photographs show the three-layered structures
of each individual cell membrane very well. Membranes appear symmetrical
in structure. (c) Molecular arrangement of these membranes. (d) Plasma
membranes after $KMnO_4$ fixation and uranyl acetate staining. The individual
plasma membranes appear to have an asymmetrical structure. (e) Molecular
arrangement of these membranes. (From L. T. Threadgold, "The Ultrastructure of the Animal Cell," Pergamon Press, 1967.)

combination of osmic or permanganate fixation with Vestopal embedding produced the same three units but they were asymmetrical.
The inner line became expanded to 35–40 A. The outer line remained at 25 A, and the space between remained at about 30 A.
The total thickness of the membrane was thus expanded from
75–80 Å to 90–95 Å. These studies also showed that the three-
layered (75-A) membrane was not equivalent to the solid line shown
in the original electron microscope studies of the membrane but
that only its 25-A inner unit was equivalent to it. The 25-A inner
unit is, in fact, equivalent to the 35- to 40-A inner unit (see Fig. 9).
The chemical nature and molecular arrangement of the membrane is
believed to be more or less the same, although there are some suggested differences. The extra width of the inner line is believed to be
due to the presence of globular protein. It is possible that some of
this globular protein may be enzymatic in nature. There is also a
possibility that the outer 25-Å layer of the membrane is not protein
but is composed of polysaccharides or possibly mucopolysaccharides. There is some evidence also that the lipid molecules of
the central layer are embedded in a crystalline hydrate matrix the
whole forming a continuum that would have important electrical and
energy properties. There is also evidence that there is a thick layer
of mucopolysaccharide attached to the outside of the cell membrane,
but that the process of fixation and embedding removes most of
it—thus its thickness cannot be ascertained at this time, although
it has been measured in ameba and found to be 200 A thick
and has extensions measuring up to 2000 Å extending from it.
What the function of this layer is cannot be established for certain,
but there is evidence that it binds large molecules that are ultimately
taken into the interior of the cell by pinocytosis.

In the apical region of cells, the plasma membrane may be
folded into fingerlike projections called microvilli. These are very
numerous in the apical borders of kidney tubule cells (3000 per
cell) and those of the small intestine (1500 per cell). These villi
are completely covered with "unit membrane."

Some of the ideas of the nature of the cell membrane had
been derived from a study of its fundamental properties. First of
all it is known that lipid-soluble substances have a preferential permeability across cell membranes, and this suggests that there must
be in the membrane a continuous layer that is composed of phos-
phatides, steroids, or fats or combinations of those substances.

Also cell membranes have a high electrical resistance that, Danielli points out, is further evidence that there must be a continuous layer of lipids. Another property is the existence of a low surface tension at the surface of the cell membranes. We have mentioned earlier that Danielli has suggested this property indicates that on the surfaces of the membrane protein layers are adsorbed. Cell membranes are disrupted by digitonin, which has a special ability to form a complex with cholesterol, and this suggests that this latter compound may play a part in the structure of the membrane.

Analysis of red cell ghosts has in fact demonstrated that a relatively large amount of cholesterol is present in the membrane. An interesting theory concerning the structure of the red cell membrane was proposed by Winkler and Bungenburg de Jong in 1941. This theory gives the free cholesterol the role of stabilizer for the charged phospholipid molecules thus enabling them to give form to the membrane. The essential features of this structure are seen in Figs. 5 and 6 (see also Fig. 10). This concept has been elaborated

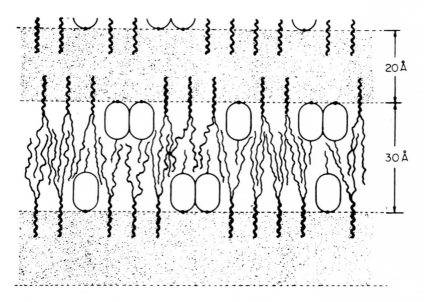

FIG. 10 Diagram of a possible arrangement of lecithin and cholesterol molecules in a bimolecular leaflet (membrane). The cholesterol shows as oval structures. (From A. M. Glauert and J. A. Lucy, Globular Micelles and Membrane Lipids, *in* "The Membranes," Academic Press, 1968.)

by Frey-Wyssling in 1953, whose studies on submicroscopic morphology and its relation to the chemical constituents of the cell are of fundamental importance. It is of interest to notice that in the two tissues where cholesterol has been assigned the major role of structural component, it is present mainly if not entirely in the free unesterified form. There is one striking difference however in the nature of the cholesterol in the myelin sheath and the red cell. In the myelin sheath it is apparently stable and once deposited under normal conditions remains static for life. In the red blood cell there is a steady turnover and exchange with plasma cholesterol. In general, in tissues where there is active metabolism of cholesterol, esterified cholesterol is found, the amount varying from tissue to tissue.

We have indicated that there is evidence of pores in the plasma membrane. Studies by Kavanau and others have shown these pores may be open or closed and since they are present in large numbers in the plasma membrane when they are open they compress the membrane into a series of pillars composed mainly of lipid: These pillars or "micelles" may be about 80 Å across and as much as 200 Å long, the pores in such cases would contain a matrix of water. The protein components of the membrane would lie on the ends of the pillars. In the closed condition of the pores, the structure of the membrane would resemble that already described. There is some evidence that the changes brought about by fixation and embedding in electron microscopy produce images of the plasma membrane in which the pores are almost invariably in the "closed" state (see Figs. 11–13).

One should, perhaps, at this point say a word about the composition of the myelin sheath since this is in a sense an extended membrane (see Fig. 6). (A more detailed description of the myelin sheath is given at the end of this book.) Finean (1953, 1956), from his studies on the x-ray diffraction pattern of myelin, suggested that it was probably composed of alternate layers of lipid and nonlipid, the nonlipid probably being protein. The lipid molecules appear to be curled or tilted so that their full length is not extended, and there is probably a stable complex formed between the cholesterol molecule and the larger phospholipid molecules. It is postulated that the free hydroxyl group of the cholesterol is important in binding the molecule by associating with the polar end of the lipid chain which it curls round. The hydrocarbon part of the cholesterol molecule is bound to the phospholipid by van der Waals forces. Cholesterol

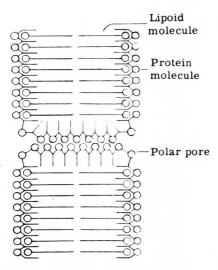

FIG. 11 Diagram of cell membrane as suggested by Danielli in 1954. (From *Colston Papers* **7**, 1, 1954.)

has a high dielectric constant, and because of this it might act as an insulating agent.

THE PERMEABILITY OF CELL MEMBRANES

Danielli has discussed on several occasions the various problems associated with the permeability of natural membranes, and the account that follows is taken largely from his writings.

Penetration of compounds

Danielli points out that there are three sites of resistance to free diffusion in cell membranes: (1) the membrane/water interface for diffusion into the membrane; (2) the membrane/water interface for diffusion from the membrane into the water; and (3) the interior of the membrane. Every molecule that needs to pass into the interior of the cell has to get through these three sites of resistance. A molecule that has polar groups (—OH groups), as opposed, for instance, to nonpolar groups such as methylene, forms at least one hydrogen bond with water for each polar group; all these hydrogen bonds must be fractured simultaneously if the molecule is to penetrate even into the lipoid membrane. Glycerol, having three OH groups, must acquire sufficient kinetic energy to break three hydrogen bonds simultaneously before it can penetrate into the membrane. This involves a large amount of energy, consequently resistance 1 (diffusion from water to membrane) is so high for glycerol that resistances 2 and 3 are reduced to insignificance. On the other hand, a molecule such as methyl alcohol, which has only one OH group and one CH_3 group, penetrates easily into the membrane and can also pass easily out of the lipoid layer into water, and presumably ethyl alcohol (C_2H_5OH) can function similarly. Thus we find that for such molecules the rate of diffusion across the interfaces is very large as compared with rate of diffusion across the membrane. The interior of the membrane in the case of these compounds is the most important factor controlling penetration. Differing from these examples are molecules that are predominantly hydrocarbon in nature such as carotene ($C_{40}H_{56}$). With a molecule such as this, resistance 1 is insignificant; the molecule has no polar groups and thus even resistance 3 is not enormous; however, resistance 2 is very large indeed, because the hydrocarbon groups are hydrophobic and a considerable amount of kinetic energy is required to transfer CH_2 groups from lipid into water. When many such groups are present, they must all be transferred simultaneously from the lipoid layer into water, since otherwise the molecule remains substantially part of the lipoid layer and cannot diffuse away into the aqueous phase.

With these examples in mind, Danielli classified penetrating molecules into four groups.

(A) Molecules with few polar and few nonpolar groups—for these resistance 3 is most important and they can penetrate comparatively rapidly, e.g., oxygen and methyl alcohol.

(B) Molecules having a predominantly polar character—for these resistance 1 is most important and penetration is slow, included are glycerols, sugar, and glycogen.

(C) Molecules having few polar and many nonpolar groups—resistance 2 is most important, penetration is slow, e.g., carotene, vitamin A, and fat.

(D) Molecules having many polar and many nonpolar groups—resistances 1 and 2 are both important and penetration is slow, e.g., polyhydroxylic bile acids, the glucuronide of oestrin, and proteins.

Danielli has pointed out that these principles of membrane permeability have a distinct physiological significance. For instance, oxygen is required in large amounts by the cell and carbon dioxide must be disposed of rapidly, and both of these substances can penetrate the cell membrane rapidly. On the other hand, the first products of glucose utilization, for example, by muscle cells are glycerol derivatives, these are valuable and the cell membrane does not let them get through too easily and so they escape only slowly. During sustained work lactic acid is formed—this would be toxic if it accumulated. The cell membrane is relatively permeable to lactic acid that thus can escape into the blood, and this permits violent exercise to be maintained for a much longer period than would be the case if the cell membrane was impermeable to lactic acid. Amino acids penetrate all membranes moderately well but protein penetrates badly and, therefore, amino acids are stored as protein. Fatty acids will penetrate moderately well, but neutral fat very poorly; therefore fatty acids are stored as neutral fat. Glucose that can get in and out of cells fairly rapidly is polymerized in liver cells to form glycogen, which passes the cell membrane only

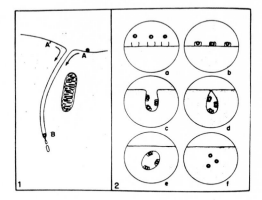

FIG. 12 Diagrammatic representation of pinocytosis after Bennett (*J. Biophys. Biochem. Cytol.* **2,** part 2, Suppl. 99, 1956). Pinocytosis is described by Holter (*Intern. Rev. Cytol.* **8,** 481, 1959) as "a mechanism for the discontinuous uptake of solutions by invagination and vesiculation of the cell surface. The quantities taken up at a single gulp cover a wide range from submicroscopic vesicles to large vesicles and there is no sharp demarcation between phagocytosis on one side and molecular permeation on the other." This is illustrated in the diagram.

with great difficulty. Each of these three latter products, to which the membranes are impermeable, are the results of polymerization of simpler compounds.

It is of interest that, when compounds that are diffusing through a cell membrane from outside to inside are being incorporated into some compound within the cell, they appear to penetrate the membrane more easily. A typical example of this can be drawn from amino acids. If they are being actively incorporated into proteins, the rate of diffusion through the membrane will be much more rapid than if they were not being incorporated. Presumably, if they are being built into proteins, they cease to accumulate and so cease to build up a concentration gradient against the passage of further amino acids.

Detoxification mechanisms also take advantage of the principles of the cell membrane permeability. For instance, toxic substances which might penetrate the cell membrane become conjugated with amino acids, sulfuric acid, or glucuronic acid. Thus toxic cell-penetrating substances such as bromobenzene or menthol are converted into new molecules that penetrate the cells with great difficulty and once in the bloodstream tend to be filtered off by the glomeruli of the kidneys and cannot be reabsorbed from the urine and returned to the bloodstream by the kidney tubules.

Penetration of ions through cell membranes

Penetration of ions through cell membranes is of special interest because for instance, Na^+ ions are more concentrated outside the cell (in the body fluids) and K^+ ions are more concentrated inside the cell, yet we know that K ions will pass into the cell and Na ions will pass out. Both these passages thus take place against a concentration gradient, and this can only occur by the application of energy. It is claimed that oxidative enzymes and phosphatases play a part in the movement of ions in and out of membranes; a process known as "active transport." Substances that interfere with the activity of these enzymes have been shown in many cases to interfere with the movement in or out of the cell of both sodium and potassium. There are certain specialized cell membranes, e.g., the cell membrane of the intestinal cells, in which the distal edge of the cell is folded to produce a larger number of extremely small fingerlike processes called microvilli; these microvilli are covered

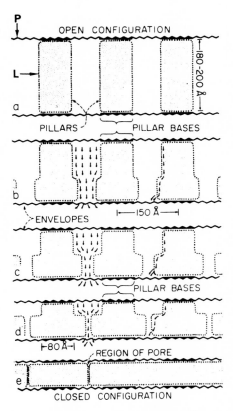

FIG. 13 Diagrams showing the configurations of the plasma membrane at various states of pore closure. (From L. T. Threadgold, "The Ultrastructure of the Animal Cell," Pergamon Press, 1967.) L, lipid; P, protein.

with a typical cell membrane. There are a large number of dephosphorylating enzymes localized around these microvilli. Their precise function in this position is not known, but it is almost certain, however, that they play a part in active transport by utilizing the energy derived from the hydrolysis of high-energy phosphates.

It is obvious from what we have said therefore, that the cell membrane, in general, is a very selective structure and exercises a good deal of control over the cell by deciding what goes in or out and by enabling the cell to be isolated from its environment. In this way it permits labor to go on in the cell that may be quite different from that going on in a neighboring cell, or in the body fluids surrounding the cell. Here then is the first division of labor in the cell to be described by us.

However, the study of the cell membrane is not yet ended. In 1952–1954, Danielli proposed the structure of the cell membrane that explained the way in which large protein molecules in the cell membrane could permit the passage of ions through the membrane. Danielli conceived of a long protein molecule extending right through the membrane and passing outside it. This end became attached to an ion and then by contraction pulled the latter through into the interior of the membrane. (See Fig. 14.)

According to Lundgard and Hodgkin, a cytochrome oxidase energy-producing system is involved in the penetration of ions. They claim that their ion moving system requires adenosine triphosphatase (ATPase), creatine phosphatase, and cytochrome oxidase. Conway has produced a redox-pump hypothesis that attempts to explain the penetration of ions through cell membranes by a mechanism that involves the successive oxidation and reduction of a compound in the membranes; possibly a phospholipid provides the energy for the movement of ions in this way. Phosphatidic acid is now thought to be concerned with ion movements through membranes.

The possible relationship of phosphatases to cell membrane permeability has also been mentioned in connection with the microvilli (brush borders) of intestinal epithelium. The passage of glucose across the membranes of gut cells could be explained by a mechanism involving these enzymes. It is of interest that in yeast cells too there is some evidence that phosphatases are localized on the membrane and that these enzymes are also involved in the permeability

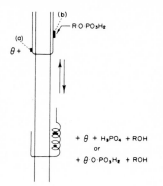

FIG. 14 Diagram of Danielli's theory of secretion by a contractile protein. (From *Soc. Exptl. Biol. Symp.* **6**, 1, 1952.)

not only to glucose but also to phosphates. Phosphatases are also present on the brush borders of the kidney tubule cells where a good deal of absorption takes place. More recent work has indicated that ribonuclease may play a part in the penetration of cell membranes by ions. Lansing and Rosenthal and Tenardo, for instance, showed that when ribonuclease was added from the outside to certain cells, it modified their permeability to ions; Brachet and Leduc found exactly the same thing for amphibian eggs. According to Brachet "amphibian eggs swell very much when they are immersed in a ribonuclease solution and since cell membranes often give strong cytochemical tests for ribonucleic acid it might well be that the integrity of this nucleic acid is of importance for normal permeability."

In another paper (1954), Danielli made a suggestion that the cell membrane consisted, as suggested before, of bimolecular leaflets of lipoid with protein molecules stretched both inside and outside the membrane, but that it also contains a series of pores that he calls polar pores (see Fig. 13). In other words, groups of protein molecules are oriented radially with their polar groups directed toward the interior of the pore and such molecules are thus able to control the penetration of the compounds or substances through this pore according to their polar group affinities.

In a later paper, Danielli (1959) discusses the association with all membranes of a group of enzymes, which are called "permeases." These compounds he says "permit either facilitated diffusion or active transfer of 'signal' molecules across the membranes." It is possible that insulin may function as a permease for cells of certain organs. To quote Danielli again "Once permease becomes incorporated in the plasma membrane the situation is transformed: specific substrates can penetrate, more permease will become available by induced synthesis, induced enzyme will appear and a whole range of further induced enzymes may appear in response to the action of the induced enzymes of the inducing substrate. Thus a transient infection with permease may potentially change drastically the state of differentiation of a cell—possibly even result in carcinogenesis." It is very likely that there will be considerable further developments in this subject of permeases, and their relation to cell-membrane permeability, and these will be awaited with interest.

A further development that should be noted when considering problems of penetration of compounds into cells is that of pinocyto-

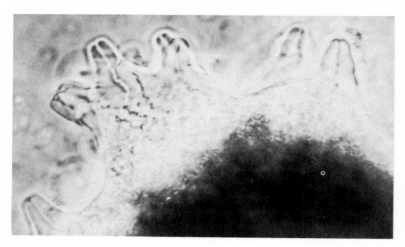

FIG. 15 Phase contrast study of pinocytoses in *Ameba proteus*. Channels (white) can be seen extending in from the edge of the cell. (Photograph by D. M. Prescott. From Holter, *Intern. Rev. Cytol.* **8,** 481, 1959.)

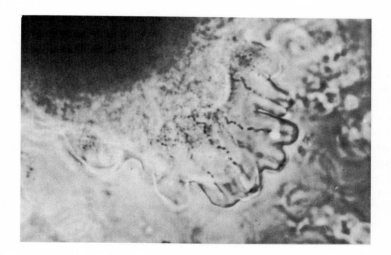

FIG. 16 Phase contrast study of pinocytosis in *Ameba proteus*. Strings of vacuoles can be seen occupying approximately the same position as the channels in Fig. 13. (Photograph by D. M. Prescott. From Holter, *Intern. Rev. Cytol.* **8,** 481, 1959.)

sis. This literally means "drinking by cells" and was first described by Warren Lewis many years ago in tissue culture cells. In this process, these cells were seen by a pseudopodial-like movement to engulf a droplet of culture fluid in which they were lying. The process was equivalent and similar to the phagocytosis of solid food by the ameba, only in the case of the tissue culture cell it is a droplet of fluid that is engulfed. Now it seems that the cells of multicellular animals even *in vivo* can do the same sort of thing though there may be specialized parts of the cell membrane which carry out the process. A case in point is the kidney tubule cell. From the basal part of the cell, long internal extensions of the cell membrane pass a considerable distance up into the interior of the cytoplasm. They are present in some other cells too and are called "caveolae intracellulares." The cell appears to be able to imbibe droplets of fluid by an engulfing movement of the membranes of these caveolae. It is possible too that the cell membranes of a variety of cells may be able to carry out pinocytosis even without the presence of these intracellular membranous extensions. (See Figs. 12, 15, and 16.)

Thus we see that, if a cell really "wants" to take in a few macromolecules that pass its membranes with difficulty, it can always take them by "the scruff of the neck" and pull them into its interior.

The actual process of taking large molecules into the interior of the cell is known as "rhopheocytosis." This differs from pinocytosis in that the latter involves a "gulping in" of globules of fluid by the cell membrane—these globules vary from 0.01 to 2.0 μ, and if large molecules are taken in with this globule, this is just fortuitous. In the case of rhopheocytosis the large molecules first became attached to the cell membrane, which is then stimulated in some way to invaginate and so carry them into the interior. There exists also a process known as "cytopemphis" or "podocytosis." In this process molecules are engulfed into the cell as described above but the vacuole containing them moves across the cell and discharges its contents onto another surface of the same cell. This process is very well seen in kidney tubule cells and intestinal epithelial cells. (See Fig. 17 for localization of enzymes on membranes.)

In many cells, free-living and those forming part of a multicellular animal or plant, projections of the cell surface occur that are known as cilia or flagellae. Although under the light microscope

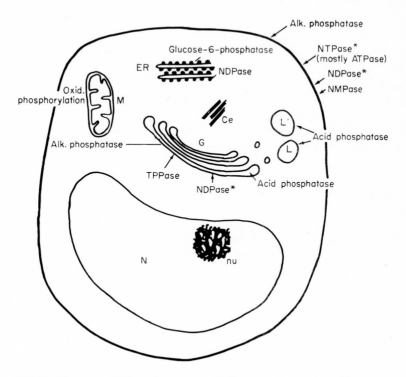

FIG. 17 Diagram of different enzyme activities associated with cellular membranes. Alk pase, alkaline phosphatase; NTpase, nucleoside triphosphatase; ATPase, adenosine triphosphatase; NDPase, nucleoside diphosphatase; NMPase, nucleoside monophosphatase; acid pase, acid phosphatase; TPPase, thiamine pyrophosphatase; N, nucleus; Nu, nucleolus; G, Golgi vacuoles; L, lysosomes; Ce, centriole; M, mitochondria; ER, endoplasmic reticulum. (From G. de Thé, Ultrastructural Cytochemistry of the Cellular Membranes, in "The Membranes," Academic Press, 1968.)

these structures appear as simple projections, the electron microscope has shown that they actually have a complex internal structure. The membrane of a typical cilium or flagella is continuous with the cell (plasma) membrane and represents an extension of it (Fig. 18). In the center of the organelle is a bundle of fibers known as the "axoneme." In the center of the axoneme are two fibers—these are surrounded regularly by nine pairs of fibers, the two members of each pair being very closely related to each other. The two central fibers are surrounded by a sheath. In each of the nine pairs of peripheral fibers one fiber is a little smaller than the other and lies nearer to the center of the cilium or flagella. Each of the fibers, central and peripheral, appears as a tube in cross section and the wall of the tube is 45 Å thick. There is little doubt that this structural configuration has a functional significance (see Fig. 19).

The enzyme adenosine triphosphatase, which is of course a protein and an energy-releasing enzyme (see below), has been found in cilia. Lack of solubility of the proteins of cilia prevented for some time a study of the chemistry and physical properties of the proteins of cilia, but I. R. Gibbons has found that in one form of protozoan (and this may be the case in the ciliated cells of higher forms) the lack of solubility was due to the membrane around the cilium and that when this was disrupted the protein of the cilium or flagella could be extracted. At least two adenosine triphosphatase proteins have been extracted this way and have been called "dynein." Differential extraction of the cilia shows that most of the enzyme protein (dynein) is located in the axoneme. These enzyme proteins have certain differences from those found in muscle,

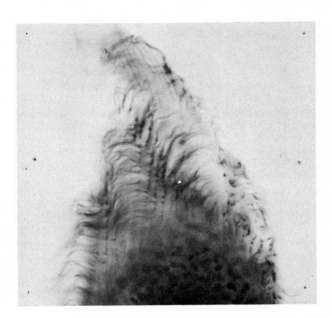

FIG. 18 Surface section of *Opalina* showing cilia, basal granules, and myoneme threads. (Photograph by K. C. Richardson.)

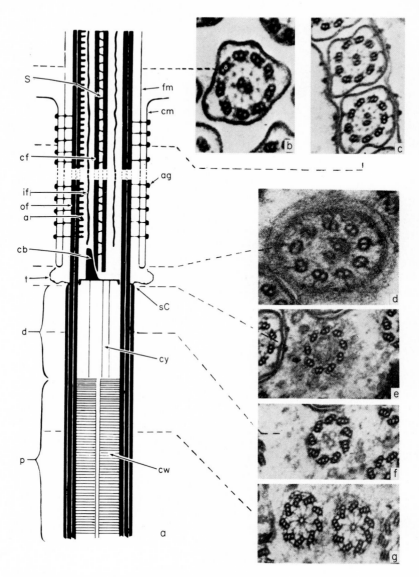

FIG. 19 Diagram of cilium of *Pseudotrichonympha* with electron micrographs at various levels. All EMs at 95,000× with the exception of d, which is 140,000×. a, arms; ag, anchor granule; bp, basal plate; cf, central fiber; cb, crescentric body; cm, cell membrane; cw, cartwheel structure; d, distal region of basal body; fm, flagellar membrane; if, inner fiber; of, outer fiber; p, proximal region of basal body; s, central sheath; sc, distal end of subfiber c; sf, secondary fiber; t, transitional fiber. (From L. T. Threadgold, "The Ultrastructure of the Animal Cell," Pergamon Press, 1967.)

but their presence emphasizes one of the fundamental properties of contractile or motile tissues. We still do not know the precise mechanics of the movement of cilia or flagella, but at least we have made a start.

DIFFERENTIAL CENTRIFUGATION

One of the developments in technique that has done an enormous amount to advance our studies of the cell is that of differential centrifugation, and an account of this technique should be given before we pass on to the study of the cell components.

The extraction of cell components from tissues dates back a very long time, and studies on the chemical nature of the nuclei were carried out during the 19th century on nuclei obtained from pus cells by differential centrifugation. More recently, tissues have been homogenized and spun down at different speeds, and the different portions of the cell separate out according to their densities. For example, if a homogenate is taken up in 0.8-M sucrose solution, which tends to preserve the morphological character of most cell structures, and it is then spun at about 1000 rpm, the cell debris and the nuclei come down. The nuclei can be differentially centrifuged again from the cell debris so that a pure nuclear fraction suitable for chemical analysis can be obtained. At a higher speed, the mitochondria and at still higher speeds, the particles of glycogen, one part of the microsomes, and other parts that are really fragmented endoplasmic reticulum aggregate at the bottom of the centrifuge tube. As a result of this technique, studies of the enzyme content of the nuclei and mitochondria have been made in great detail. This has caused some controversy because of the possibility that soluble enzymes and other substances might be leached from the nuclei during the homogenization procedures. Dilute acids, e.g., citric and acetic acids, have been used as homogenization fluids and possess certain advantages in that they make the tissue fragile and thus more easily broken. They also appear to harden the nuclear membrane and help to preserve the nuclei intact. However, they are unsuitable for enzymatic studies, and tissue homogenized in solutions of sucrose are usually used for this purpose. The latter method is claimed, because of the low ionic strength of the sucrose solution, to reduce greatly the loss of materials from the nuclei. It is said, in fact, that proteins and nucleic acids do not get through the membranes but low molecular weight substances do suffer some loss by this method. Allfrey and his colleagues at The Rockefeller University have tried to reduce any type of loss from the nucleus by freeze-drying the tissue and then grinding and floating it in non-aqueous media. Although in many respects this is advantageous, it appears difficult to be certain that small fragments of cytoplasm do not stick to the nuclei and contaminate the material. Extremely interesting results have been obtained by this particular technique.

Another check that was made on this differential centrifugation technique was to add various pure enzymes to the homogenate and then to calculate how much was adsorbed on the nuclei, mito-chondria, microsomes, and so on and how much was left in the supernatant. Although the amounts taken up with the supernatant varied with different tissues, generally speaking, the results were encouraging enough to indicate that the adsorption of enzymes from the normal supernatant was not really a significant cause of error in this technique.

Roodyn in a review of the enzymic content of isolated nuclei has discussed the possibilities of contamination. He points out that we know that there is a very intimate connection between the endoplasmic reticulum and the nuclear membrane and that it is believed that the nuclear membrane is really a fold of the endoplasmic reticulum. Thus the nuclear fraction could easily be contaminated with this part of the cell. Also there is a possibility that some of the sedimented material may contain whole cells that have escaped homogenization as well as nuclei, and this would be another source of error. There is a possibility too that mitochondria may stick to the isolated nuclei and so also be contaminants. Al-though a certain amount of contamination is thus possible, it is only significant where a small percentage of total enzyme content of the cell is attributed to the nucleus, in this case it may be due to con-tamination, but where the nucleus contains an appreciable pro-portion of the total enzyme activity of the cell then the contamination becomes not very significant.

THREE
The
cytoplasm

ITHIN the cell membrane and separated from the outside (extracellular) world by it, is the cytoplasm. Does it perform any labor? What is its contribution to the life and welfare of the cell? First of all let us find out what we can about its nature.

Cytoplasm chemically is composed of proteins, lipoids (which include fatty, phospholipid, and steroidal compounds), carbohydrates, mineral salts, and, of course, a good deal of water (in most cells more than 90%). The constituents of the protoplasm of animals and plants are generally similar, although the relative proportions of the various components varies a good deal in different organisms and probably even in the same organism under different physiological conditions. Originally, according to Wilson in his classic work "The Cell" (1928), the similarity between the protoplasm of plants and animals was stressed and particularly the importance of protein as a structural element in both of them (this, of course, can still be accepted). However, the chemical evidence has since shown that some of the other elements differ in animals and plants, for instance, the plant protoplasm has rather more carbohydrate in it, and proteins and lipids are the main constituents of the animal protoplasm. Protoplasm, in general, behaves and appears to be something in the nature of a colloidal system that is very complex and behaves almost always as if it were a viscous liquid. This is particularly well demonstrated by living protoplasm undergoing streaming movements so well demonstrated in plant cells and in the cells of animals, such as *Ameba,* when pseudopodia are being produced. Cells that are lying free tend to round up and become spherical when they are resting, and if small fragments of protoplasm are chopped off from cells by a process described technically as "micrurgy," which simply means micro-surgery, the little pieces so removed become spherical. Both these facts are in keeping with the conception that protoplasm is a viscous fluid.

Viscosity varies a great deal in different kinds of cells and it varies in the same cell from time to time according to the physiological state of the cell. Sometimes it may set into a semi-jellylike condition, and this is particularly well shown by some types of slime molds (myxomycetes). When one of these is touched the whole organism suddenly sets into a gel formation, and only after the passage of time does the organism gradually pass into a viscid fluid state again.

The ground substance of the protoplasm is known as hyaloplasm and contains a number of bodies and structures, vacuoles, and so on and also a number of very tiny particles, some of them ranging into the ultramicroscopic level and which undergo active Brownian movement. According to Wilson "Flemming observed the dance of minute fat drops in living cartilage cells." Many of the theories of the nature of protoplasm developed in the latter part of the 19th century. At first there were not many theories about its structure because living cytoplasm is optically homogeneous or empty; and between the years 1870 and 1890, when great interest was taken in examinations of fixed and stained tissues, a number of theories of the structure of protoplasm were developed based on its appearance after various types of fixation. Probably one of the most well-known of these is the fibrillar theory. This held that the protoplasm was made up fundamentally of delicate fibrils that were either separate strands or formed a meshwork and this was situated within a homogeneous, optically empty, ground substance. Among the people whose names were associated with this theory were Leidig, Flemming, Carnoy, and Heidenhain. These fibrils were thought to be of fundamental importance to the vital activities of the cell. However, later on, the fibrillar theory became subdivided into two subsidiary theories. One was the reticular theory and the other the filar theory.

The reticular theory was a modification of the filar theory in the sense that, whereas the fibrils did not form networks in the filar theory, they did in the reticular theory. However, toward the end of the century a number of workers including Flemming, Bütschli, and Fischer showed that coagulation artifacts could produce all the phenomena that had been described as fibrillar structures in the cytoplasm. A number of experiments had been carried out by various workers in which every conceivable type of fibrillar structure that had been described in the cell had been duplicated in nonliving material made up, for instance, of white of egg or gelatine that had been coagulated by various fixatives. Thus by the end of the 19th century, the fibrillar theory and its offshoots were beginning to fall into disrepute.

Another theory of the structure of protoplasm was the alveolar or foam theory (see Fig. 20) which was enunciated by Bütschli in a series of papers, the first of which he published in 1878. He believed that protoplasm was made up of a series of what he described as "alveolar spheres" that were suspended in a hyaloplasmic material. He regarded protoplasm as being equivalent to two viscid liquids, one of them forming the walls of the spheres scattered among the continuous substance (the hyaloplasm). There were also present in the cytoplasm a number of very small granules described as "microsomes," and we should note that the term "microsomes" as used by the early cytologists is quite different from the use

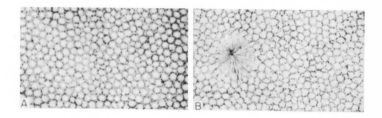

FIG. 20 The structure of cytoplasm. The alveolar theory. (A) Appearance of living cytoplasm of the starfish (*Asterias*); (B) appearance of the cytoplasma after fixation with sublimate acetic. (From E. B. Wilson, "The Cell in Development and Heredity," Macmillan, 1928.)

of the term today. Nowadays "microsomes" have a fairly specific connotation since they are those bodies that are spun down in an ultracentrifuge from cell homogenates; the last of all the particles of the cell to be spun out. To what extent all modern microsomes represent something that really occurs in the living cell or not we shall discuss later on.

Another theory which developed in the 19th century was the "granule" theory. This was introduced by Altmann in three papers in 1886, 1890, and 1894 and was subsequently developed by quite a number of other writers, mainly Benda and Meves. Altmann, using a special technique, demonstrated red-staining fuchsinophil granules in all the cells he examined. In many cases they were a fairly uniform size, and in some cells they were very closely crowded and appeared to occupy nearly all the cytoplasm. Altmann regarded these granules as elementary organisms and he called them "cytoblasts" or "bioblasts' and believed that they "lived" in a homogeneous ground substance again called the "hyaloplasm." We know now, of course, that Altmann's granules were actually mitochondria. (See Fig. 21.)

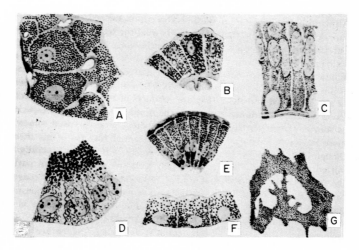

FIG. 21 Granular structure of cytoplasm as demonstrated by Altmann. We know now these are mostly mitochondria. (A) Liver of mouse, (B) tubules of mesonephros, embryo chick, (C) intestinal epithelium of frog, (D) pancreas of *Triton* showing secretory granules and fibrils, (E) epithelium of intestinal villus, cat, (F) Harderian gland of rat, (G) small portion of pigment cell with pigment granules, salamander larva. (From E. B. Wilson, "The Cell in Development and Heredity," Macmillan, 1928.)

Wilson, in 1928, summing up the theories of cytoplasmic structure, said, "We are driven by a hundred reasons to conclude that the protoplasm has an organization that is perfectly definite but it is one that finds visible expression in a protean variety of structures. We are not in a position to regard any of these as universally diagnostic of the living substance." He went on to say that the fundamental structure of protoplasm lay beyond the then limits of microscopic vision. His final conclusion was that protoplasm possessed an ultrastructural configuration known as a "metastructure."

The attempts to demonstrate definite structure in protoplasm were followed by a reaction against the alveolar, fibrillar, and granular theories and led back again to the reverse conception that the material was structureless. Yet during this pre phase-contrast, pre-electron-microscope period, some prophets were found amongst the biochemists who are, or at any rate *were* not noteworthy for having a morphological outlook on cells. These prophets demanded the presence of a cytoskeleton, which they felt was a necessary structure along which various enzymes, etc., should become aligned. One of the first steps that helped to bring back the conception that structure existed in apparent hyaloplasmic cytoplasm was the development of the phase-contrast microscope in which a number of formed structures were seen to exist in the cytoplasm of the living cell.

Immediately beneath the cell membrane, the protoplasm of the cell is differentiated to form an "ectoplasm," which appears to be in a partly gelled condition. It is probable that this ectoplasm plays an important part in the movements of the cell, and it appears to be actively forming and reforming in the movement of protocells such as the ameba. It is also modified to form cilia. These are, in effect, ectoplasmic prolongations surrounded by a cell membrane. The ectoplasm is also a constituent part of the microvilli, which occur on absorbent cells of the kidney tubules and intestine.

Phase-contrast microscopy demonstrated a good deal of structure in the apparent hyaloplasm, and later this was supported by the results obtained with a further development of this sort of microscopy—the interference microscope in which objects of different density in the cell were colored differently according to their density. Within the last twenty years the electron microscope, using the fixation by osmium tetroxide, and other compounds, has demonstrated an incredible complexity in what was only a few years ago thought of as a relatively homogeneous hyaloplasm.

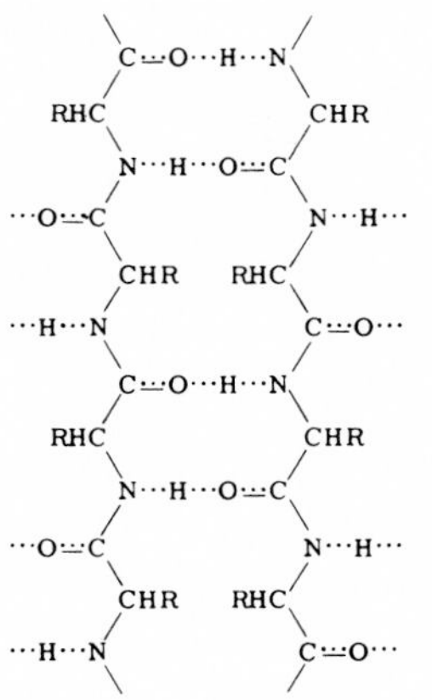

$$C{=}O{\cdots}H{\cdots}N$$

FIG. 22 Hydrogen bonds between polypeptide chains. (From Frey-Wyssling, "Submicroscopic Morphology of Protoplasm and Its Derivatives," Elsevier, 1948.)

The cytoplasm is filled with mitochondria, microbodies of various sorts, double membranes, Golgi apparatus, and various granules; in fact, there is relatively little of what we might describe as hyaloplasm. This hyaloplasm does, however, exist, and when fixed and examined under the electron microscope it seems to be composed of an extremely fine network, which becomes coarser or finer according to the type of fixative used. Now what is this network composed of? Wykoff has pointed out that a number of proteins that contain long, filamentous-type molecules (see Figs. 9, 22, and 23) are capable of forming gels that have many of the physicochemical properties characteristic of protoplasm. He points out that gelatin is one of these and describes one of his experiments in which gelatin gels were first fixed with osmic acid and then examined under the elec-

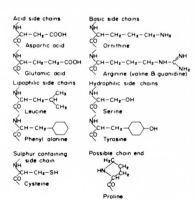

FIG. 23 Side chains "R" of the polypeptide chains. (From Frey-Wyssling, "Submicroscopic Morphology of Protoplasm and Its Derivatives," Elsevier, 1948.)

tron microscope. He found that the gelatin was arranged in a network and that the pores of the net were small or large according to the concentration of the gelatin/gel or the fixative used. With a concentration of 2 to 4% in a gel, the net that results possesses a pore size that is very similar to that found in cells, but the pores get smaller if the gel is more concentrated. It is probable therefore that the network-like structure of the hyaloplasm we see is really to some extent the effect of the fixative, although it is partly the expression of the nature and distribution of the filamentous molecules of protein that form the basis of the hyaloplasm. These molecules are, in fact, probably arranged in a sort of network, a structure that Sir Rudloph Peters once described as the "cytoskeleton," which he thought was the basis of cytoplasmic structure. This suggestion was made some years before the electron microscope demonstrated such a wealth of formed components within the hyaloplasm itself.

Within the network of filamentous molecules (both protein and polysaccharide) that build up the hyaloplasm, a variety of molecules, not only free protein, polysaccharide, and lipid but also ions, move about attaching and detaching themselves to the network in accordance with the electrical and other forces operating at that level (see Fig. 23). We can, indeed, picture this hyaloplasmic network as a sort of skeleton for the rest of the cell.

Most scientists believe that the water in the cell is composed of unoriented molecules, distributed in random fashion as they are in a drop of water. However, three groups of workers have now produced evidence that this is not so and that the molecules are arranged in a crystalline fashion similar to that which they occupy in ice. This theory has important implications for cell and nerve physiology. The explanation, for example, of the mechanism of nerve conduction depends upon the accepted view that sodium and potassium ions are able to migrate without hindrance in the liquid water of the cell. The new theory, however (*Scientific Research*, September 1, 1969, p. 39), suggests that "intracellular" sodium and potassium ions are complexed to sites on macromolecules, and these ions hop from site to site through an ice-like matrix, obeying laws analogous to those governing the conduction of electrons in semiconducting solids.

Now in this watery or ice-like structure (whichever the case may be) lie the rest of the cell elements that have to be separated from it to carry out the types of activity in which they specialize.

THE ENDOPLASMIC RETICULUM

A prominent feature of the cytoplasm of cells, particularly pancreatic cells, under the electron microscope, is a basophilic fibrillar structure, which takes us back a little into the fibrillar theory of protoplasm; this structure is known either as the "endoplasmic reticulum" or "ergastoplasm," and reference will be made again to these names later on.

The structure and function of this material has been worked out and recognized in the last few years with the aid of the electron microscope, although it was originally discovered a long time ago with the light microscope. The discoverer was Garnier and the year 1897. The material discovered was a basophilic fibrillar material that could be seen in stained cells, particularly in the basal region of gland cells, and it was called by Garnier "ergastoplasm."

The ergastoplasm appears to be part of the system described as the endoplasmic reticulum. This latter structure was described by Porter and Thompson in a paper in 1947 on electron-microscope studies of a chick macrophage (see Fig. 24). The macrophage, when

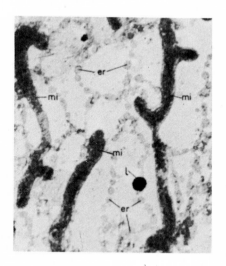

FIG. 24 Endoplasmic reticulum and mitochondria in tissue culture cells (chick macrophage). er, Endoplasmic reticulum; mi, mitochondria; l, lipid. (Preparation and photograph by Dr. K. Porter. From De Robertis, Nowinski, and Saez, "General Cytology," Saunders, 1960.)

very thinly spread out, showed that the cytoplasm was everywhere permeated by a tenuous network that they described as a lacelike reticulum. This they quite understandably called an "endoplasmic" reticulum. Later on, when fine sectioning of cells became possible in the 1950's, the endoplasmic reticulum seemed to occupy the position that was normally occupied by the "ergastoplasm," as Garnier had called it. In addition it also appeared to spread through the other parts of the cytoplasm of the cell and to be present, to a greater or lesser extent, in practically all the cells that have been examined. However, at this point we should try to solve the problem of terminology. One of the characteristics of ergastoplasm is that it is basophilic, but not all the endoplasmic reticulum is basophilic although both basophilic and nonbasophilic regions in ultrathin sections of cells have the same double membrane structure. The difference is that the basophilic portions have ribonucleoprotein particles attached to the membranes and the nonbasophilic parts do not. Therefore the ergastoplasm can be thought of as a specialized basophilic part of the endoplasmic reticulum. Sjöstrand tried to avoid this controversy by giving the names of Greek letters to various membranes in the cells. The membranes that form the endoplasmic reticulum he has described as the α-cytomembrane, the membranes of the mitochondria and of the cytoplasm receiving other Greek letters to designate them. Sjöstrand's recommendation for nomenclature, however, seems to be a rather cumbersome method of describing these structures.

The use of the term *ergastoplasm* for the system of cytoplasmic membranes is largely supported by the French school recently headed by the late Charles Oberlin, whereas in the United States the accent is more on endoplasmic reticulum with ergastoplasm as a specialized part of it. This is the nomenclature followed by Porter and Palade.

Since the nature of the endoplasmic reticulum and, indeed, of many other cell structures, is based on electron-microscope studies, we might perhaps briefly consider the significance of electron-microscope pictures, in general, and attempt to assess to what extent they represent a real structural condition in the living cell.

The late Dr. Oberlin has listed six points that suggest that the electron-microscope picture is a true picture of the living cell.

1. The types of structures that are seen under the electron microscope have also been seen in living cells, and they have been photographed or filmed by the phase-contrast microscope.

2. Whenever the possibility has existed for comparison of the same cells, for instance, alive as viewed under the phase-contrast microscope or in fixed condition with the electron microscope, there has been perfect agreement in the picture.

3. The same methods of fixation and observation reveal all the structures side by side and present in all cells, even in those that are very different from both the phylogenetic and the functional points of view. The appearance of these structures depends to a large extent on the perfect preservation of cells by first-class fixation.

4. In order to attain excellent pictures the cells must be fixed in the living state. This proves the great sensitivity of the observed structures and the reliability of the procedures in detecting such structural changes that take place immediately after vital functions have ceased.

5. The techniques of homogenization, fractionation, and ultracentrifugation have given the opportunity to isolate these structures and obtain them in a relatively pure condition in sufficient amounts to permit biochemical investigations, thus bringing closer the collaboration between the morphological and biochemical studies of the cell.

6. These structures, as they appear under the electron microscope, do not always produce the same aspect, but they vary according to the evolutionary phases and also differ in pathological conditions from which the cell may have been suffering.

These points are very strong ones, and possibly the strongest of all from the point of view of the very delicate structure that is demonstrated in electron microscopy is point 4. With regard to the first three, very few people have ever doubted that, for instance, mitochondria (as seen with the light microscope) existed in living cells. They have been seen in living as well as in fixed cells examined by the optical microscope, and it was no surprise to find that they were also present in electron-microscope pictures. The point that has been really at issue is whether the extremely fine structures that the electron microscopists demonstrate do really exist in life. The very fine, double-membrane structures and so on, could possibly be produced as a result of the technique used, although in view of the widespread constancy of findings this is unlikely. Nevertheless we should still keep a healthy attitude of caution in accepting all these fine structural details and be prepared to alter our views should evidence accumulate that there are errors in this depiction of ultrafine structure. At the moment, however, it appears that the electron microscope is telling near the truth so far as we can interpret it, but that it is not yet showing everything that is in the cell.

Garnier, the discoverer of ergastoplasm, brought out some very important points concerning its nature and these have been listed in an excellent article by Haguenau [*Intern. Rev. Cytol. 7* (1958)] to which the reader is referred for further details of the ergastoplasm. Garnier said that, in the basal region of all cells, without exception, there was a portion of the cytoplasm that seemed to have a fibrillar or rodlike structure, that these filaments or rods were not separate structures but were actually part of the cytoplasm and were in direct continuity with it. Haguenau points out that the electron microscope has provided an excellent confirmation of this. Garnier also said that these filaments were stained by basic dyes—those used were safranin, gentian violet, and toluidine blue—and that the intensity with which this basic staining occurred varied with the stage of secretion in gland cells. Another point he made was that the ergastoplasm was not a permanent structure and that its development was related directly to the state of activity of the cell. He said, for instance, that in gland cells the filaments appeared much more numerous when the cell, having already gone through a cycle of secretion and excretion, was preparing a new cycle. When the cell became loaded again with secretory granules, the filaments became less obvious and disappeared. His final point was that the filaments were closely related from a topographical point of view with the nucleus, that masses of the material formed laterally on the sides of the nucleus, and that sometimes the latter was completely encircled by the ergastoplasm. Garnier also thought that nuclear sap or chromatic substance originating from the nucleolus was able to pass through the nuclear membrane and enter into association with the ergastoplasm. This was actually a very significant comment, as will be seen later.

Following Garnier's work many other authors figured and described and categorized the ergastoplasm. Prenant in 1898 wrote a review on the ergastoplasm that he called the "protoplasme supérieur." He described the ergastoplasm as a very important zone of the cytoplasm that was capable of differentiating into specific structures and among these, according to Haguenau, were included the "Nebenkern," the "Dotterkern" of the germ cell and the ergastoplasm of the gland cell and even the Nissl bodies of nerve cells. These were rather interesting conclusions since they have now been supported by electron-microscope studies. Haguenau has pointed out that in the history of most discoveries there is a period in which it is first described, then a lot of other people describe it, and finally there is a period when everybody believes it is an artifact due to fixation. She points out that at about the same time as the ergastoplasm was discovered Altmann discovered mitochondria, but comments made on the latter by Benda in 1898 and 1899 served to complicate the proper interpretation and acceptance of the ergastoplasm. Haguenau divided the post-Garnier workers into three groups: the first group were represented by Morelle in 1927, who claimed that the ergastoplasm was nothing more than modified ground cytoplasm that took up basic stains, this being due not to any difference in structure but simply to a chemical difference. He claimed that this area of the cytoplasm was never really fibrillar. Then there was a second group who thought that the ergastoplasm was composed only of mitochondria that were distorted in shape. Champy in 1911, for example, stated that mitochondria and ergastoplasm were one and the same and that the preparations that showed ergastoplasm were simply preparations with poorer fixation than those that showed mitochondria. To some extent there is some justification for this point of view, since mitochondria and ergastoplasm are so closely related to each other topographically that it is natural enough for a mitochondrial stain to demonstrate concentrations of mitochondria in the same site as the ergastoplasm. The third group agreed with Garnier that mitochondria and ergastoplasm were quite different structures but these, according to Haguenau, had to fight very hard to prove it. Prominent among these was Regaud.

The critical experiment that Regaud published in 1908 was one in which he showed that, if he had acetic acid in his fixative, there were no mitochondria in the preparation but the ergastoplasm appeared quite normal. On the other hand, with acetic acid absent mitochondria could be demonstrated very conveniently but the ergastoplasm did not take up its normal appearance. Haguenau quotes Regaud (1909), "It is possible that the ergastoplasm consists of a protoplasmic support impregnated with chromatin or a closely related substance." Rees in 1940, using the polarizing microscope, made a study of the ergastoplasm and came to the conclusion that there was a homogeneous filamentous structure present in the cytoplasm of living cells, but it was the electron microscope that finally confirmed and helped to delineate the structure of the ergastoplasm

Preliminary studies by Porter and colleagues leading to the designation of this material as endoplasmic reticulum have already been mentioned. In 1950, Hillier was probably the first person to demonstrate that, in fine sections of liver, fine fibrous matter is present in the cytoplasm of the cell, but he was not able to establish its identity. Dalton in the same year also produced electron micrographs that demonstrated quite clearly that the cytoplasm contained a number of filamentous units and that these tended to be grouped in particular areas and were reduced following fasting of the animal.

The work of the French and American schools, together with contributions by the Swedish school headed by Sjöstrand, have now established a great deal of interesting information about the structure, nature, identity, and distribution of the endoplasmic reticulum and its specialized part, the ergastoplasm.

The endoplasmic reticulum appears fibrous in nature in sections and it is of interest that these fibers appear to form pairs, and they have been described, in fact, as paired membranes (see Fig. 25).

Actually, the endoplasmic reticulum takes the form of an irregular network of tubules that branch and anastomose. These tubules vary between 400 and 700 Å in diameter. In parts the tubules may be expanded into "cisternae" that may be over 1000 Å in diameter, these are simply flattened sacs. Sometimes the tubules are expanded into more or less circular "vesicles." The nature of the endoplasmic reticulum varies considerably in the cells of different physiological conditions in the same cell. Because of its predominantly tubular or flat saccular nature, sections through the endoplasmic reticulum observed under the electron microscope appear as "pairs" of membranes. Where small particles known as "ribosomes" (containing ribonucleic acid) are attached to the outside of each of the membranes of a membrane pair—the endoplasmic reticulum is known as "rough" endoplasmic reticulum (ergastoplasm); in membrane pairs where the ribosomes are absent, the reticulum is referred to "smooth" endoplasmic reticulum. Parts of such a reticulum may be 1000 Å in diameter. The rough endoplasmic reticulum is associated especially with cells that are actively concerned with protein synthesis and the latter with cells concerned with carbohydrate or steroid synthesis (see Fig. 25).

The ribosomes may be arranged on the outside surface of the

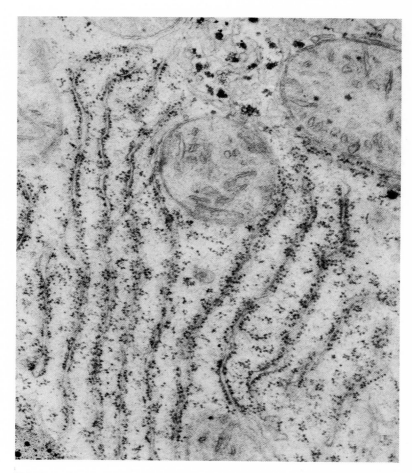

FIG. 25 ("Division of Labor in Cells," 1st ed., Fig. 17) Endoplasmic reticulum of liver cell showing double-membrane structure and RNA granules on the outside of the membrane pairs. Transverse sections of mitochondria showing fine structure can be seen. (Preparation and photograph by M. Sheridan.)

endoplasmic reticulum membrane in rows or rosettes or many other types of confirmation.

The agranular or "smooth" endoplasmic reticulum may form a very complex network of tubules extending all through the cell. Probably the most outstanding example of this is the interstitial cell of Leydig in the testis. The endoplasmic reticulum of striated muscle (known as the "sarcoplasmic" reticulum) is of the "smooth" type. The liver cell contains both "rough" and "smooth" endoplasmic reticulum—which is understandable since it is a protein synthesizer as well as a carbohydrate synthesizer. Either type of reticulum can be made to predominate in the liver cell by various drugs or by physiological demands that stimulate either protein or carbohydrate synthesis.

It is of interest that these membranes* show the 75- to 80-Å unit structure that is so characteristic of cell membranes and that may be significant in the light of something we will say later on about them. There is some discrepancy between the findings of electron microscopists regarding the ergastoplasm and the findings of the classical cytologists. The latter claim that the basophilic staining with which these fibrillar structures are associated appears to be localized at the basal end of the cell, whereas with the electron microscope these fibers or membranes extend all through the cell. This, however, can be explained in part by the fact that it is only when there are quite a number of membranes close together that they show up as a strongly basophilic area under the light microscope, and in other parts of the cell the membranes are more widely spread out and are not organized in dense groups; therefore the characteristic staining reaction is spread out over a great area of the cytoplasm and becomes diluted. Furthermore the smooth endoplasmic reticulum does not have ribonucleic acid granules attached to it (the component which produces the basophilic staining). It is also evident that some endoplasmic reticulum membranes may show as basophilic structures but do not have any obvious RNA granules—so, perhaps, this material can be associated with the endoplasmic reticulum in a molecular as well as a particulate form.

In areas where the ergastoplasmic membranes are concentrated together so that the whole cytoplasm has a lamellar appearance, it has been suggested that the term *organized ergastoplasm* should be used. This term was suggested by Houdson and Ham in 1955. Another interesting organization of the ergastoplasm is the "Nebenkern." Nebenkern are structures that are basophilic in nature. Their significance and method of formation was unknown to older cytologists, but under the electron microscope it appears that they are concentric layers of ergastoplasmic membranes that in section look

---

* However, unlike the plasma membrane, they are symmetrical in their ultrastructure.

like an onion bulb or, as some authors put it, a fingerprint. It is possible that these structures may originate from mitochondria. The ergastoplasm may also be modified in nerve cells in the region where Nissl bodies are found, as demonstrated by Palay and Palade. There are considerable variations in amount of ergastoplasm that is thought to be related to the differentiation of cells, and it has been suggested that it may also be linked to growth, and as mentioned earlier, they are particularly obvious in cells that are engaged in the production of protein.

We have already noted the association with the cytoplasm of a large number of granules (ribosomes) (see Fig. 25). It is of interest that these granules are associated with the outer part of the membrane. The side of the membrane directed toward the cavity that the membrane lines in the cytoplasm is smooth and has no granules attached to it, whereas on the cytoplasmic side of the membrane, quite distinct granules are found. They are basophilic in nature, range from approximately 120 to 150 Å in diameter, and have been the subject of considerable specuation. Since they are basophilic in nature, it is obvious that they are, as mentioned earlier, the cause of the basophilic staining of the ergastoplasm recorded by the older cytologists. The studies by Palade and Siekewitz over the last five or six years have demonstrated the very interesting fact that these granules are composed of ribonucleic acid combined with protein. The way in which Palade and Siekewitz obtained this information is of interest. They treated homogenates of cells with deoxycholate, a surface-tension reducing agent that detached the granules from the membranes. Then they were able, by differential centrifugation, to isolate a number of them, carry out chemical analyses, and so establish that they were largely composed of ribonucleic ucid (see Fig. 29). However, one should not think that all the ribonuleic acid in the cytoplasm is associated with the endoplasmic reticulum. It is present in other parts of the cell as well, but there is some evidence that about 25% of the cytoplasmic ribonucleic acid is, in fact, associated with the endoplasmic reticulum. For other illustrations of endoplasmic reticulum in different cells see Figs. 26–28.

One of the interesting observations about these particles, demonstrating perhaps not so much division of labor in cells as cooperation of labor between cell constituents, is that in the nucleolus, ribonucleic acid particles are found which are the same size (150

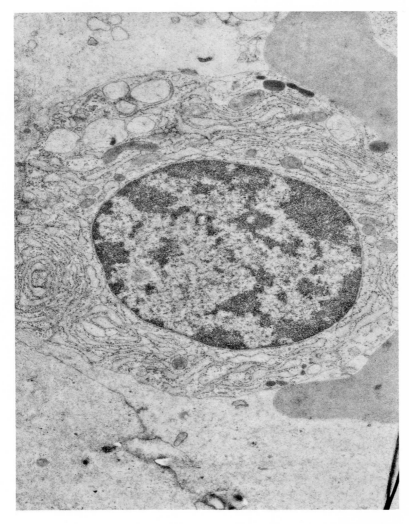

FIG. 26 Electron micrograph of plasma cell of orangutan magnified 21,000×
showing rich endoplasmic reticulum. These cells are protein producers.
(Photograph by C. Webb.)

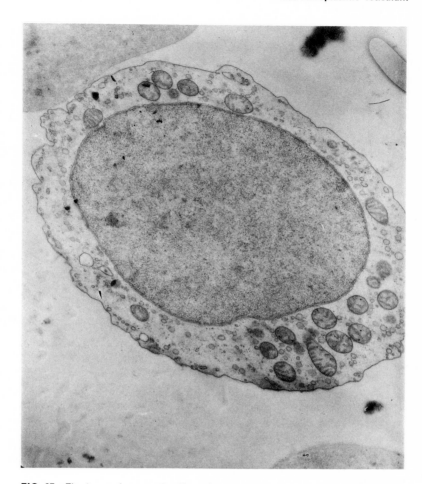

FIG. 27 Electron micrograph of lymphocyte of orangutan. Note well-defined double-membrane structure of nuclear membrane. Numerous mitochondria in cytoplasm, negligible endoplasmic reticulum. These cells do not synthesize significant amounts of protein but may develop endoplasmic reticulum and become plasma cells and secrete protein (antibodies). (Photograph by C. Webb, magnification 24,000×.)

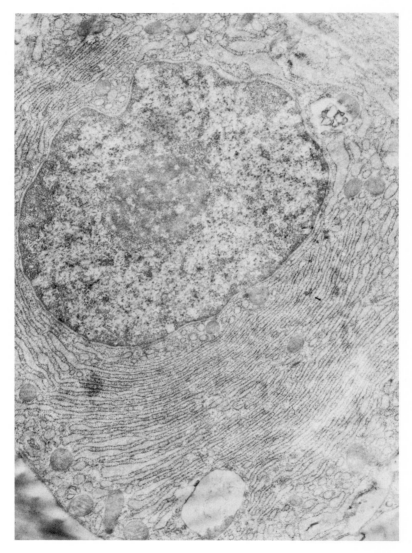

FIG. 28 Pancreatic acinar cell from Rhesus monkey, showing enormous development of the endoplasmic reticulum. This type of cell is a very potent protein producer. (Photograph by R. Tyler, magnification 21,500×.)

Å in diameter, as those associated with the endoplasmic reticulum (see Figs. 29 and 30). (It is of interest, also, that the deoxyribonucleic acid particles that are found in the chromatin of the nucleus are also about 150 Å in diameter.) It seems possible that the RNA particles could pass out from the nucleolus through pores present in the nuclear membranes (which we will discuss later) and become attached to the walls of the endoplasmic reticulum. The fact that there is a coincidence in size is of very great interest from this point of view. A different type of migration has also been suggested by Gay, who showed with the electron microscope that blebs forming at the surface of nuclei might become converted into endoplasmic reticulum and that RNA granules appeared to become attached to them before they actually became detached from the nucleus. These RNA particles or ribosomes actually contain about 40–60% of the RNA and practically no lipid material, the balance of the particle being made up of protein. The ribosomes have been called by a large variety of names including "microsomes," "microsomal particles," "Palade particles," etc. (see Fig. 29). When they congregate together as masses composed of many separate particles, they appear to become protein synthesizing units, sometimes called "polysomes" or "ergosomes" (see Fig. 30). Ribosomes occur universally in microorganisms, in plant cells, and in animal cells. In microorganisms they constitute a substantial proportion of the cytoplasm. In some primitive animal cells, the ribosomes are free in cells that are proliferating, but in cells that have differentiated they are attached to the membranes. In mammals only the mature red cell is free of ribosomes. The exocrine gland cells of the pancreas probably contain as high a concentration of ribosomes as any cell in the body. In general, cells that are actively producing protein have a high concentration of these bodies. The relationship of the E.R. and the ribosomes to protein synthesis will be discussed in the chapter on the nucleus and nucleic acids. (see Figs. 26–28 for relationship of complexity of E.R. to cells which are protein producers.)

THE NATURE OF MICROSOMES*

When Claude applied Bensley's technique of differential centrifugation to cells, isolated different cellular components, and subjected

---

* The word *microsomes* now has a different and very special application. See *peroxisome*.

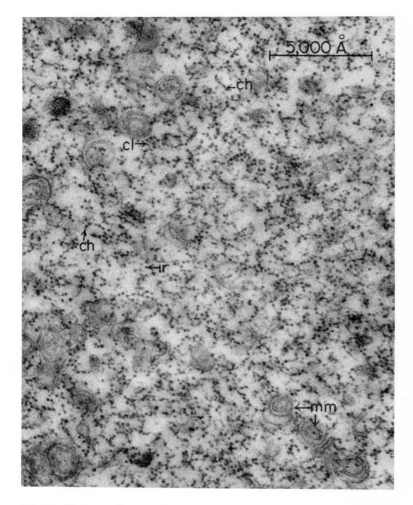

FIG. 29 Electron micrograph of particle fraction isolated from rat liver micro-somes treated with 0.26% deoxycholate. Ribosomes isolated or in chains (Ch.) or clusters (Cl.) sometimes connected by fine filaments. Microsomal membranes disposed in concentric arrays are contamination of preparation. (From M. L. Petermann, "The Physical and Chemical Properties of Ribo-somes," Elsevier, 1964.)

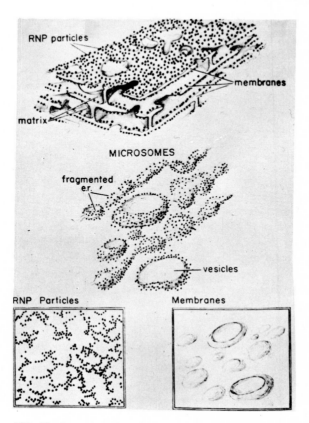

FIG. 30 Endoplasmic reticulum showing relationship of ribonucleoprotein particles to the E.R. membranes. The other illustrations demonstrate the nature of the microsomes derived from the fragmentation of the endoplasmic reticulum. Treatment of these microsomes separates the two constituents demonstrated in the lower two figures. (From De Robertis, Nowinski, and Saez, "General Cytology," Saunders, 1960.)

them to chemical analysis, he found a series of minute bodies that are submicroscopic in size, i.e., below the limit of the resolution by the optical microscope. He described these bodies as microsomes. They are the smallest bodies in cellular homogenates and come down after everything has been centrifuged off, i.e., at the highest speed of centrifugation leaving only the supernatant. Many chemical studies were done on the microsomes and a good deal of information on their enzymes and chemical composition was obtained. However, one thing that made cytologists and particularly electron microscopists uneasy was the fact that, if one looked at a liver cell (which Claude used largely as the source of material for the production of microsomes) under the electron microscope, nothing could be seen in the cytoplasm that bore any resemblance to the microsomes. Since these were of a size that could be seen at the magnification used by the electron microscope, it was difficult to understand where they came from in homogenates. The microsomes contain a good deal of ribonucleic acid and so does the ergastoplasm. Finally, by isolating the pellet of microsomes centrifuged out from the homogenized cells and by examining ultrathin sections of them under the electron microscope, various workers showed that the microsomes were small vesicles with ribonucleic acid granules attached to them. It thus seemed quite obvious that the grinding up of the cell during homogenization broke up the endoplasmic reticulum and the bits formed themselves into little vesicles that floated in the fluid and were finally centrifuged down. This work was first done by Slatterbach in 1953 and subsequently by Palade and his colleagues.

Relation of endoplasmic reticulum membranes to cell membrane

The E.R. (endoplasmic reticulum) membranes have a structure (as mentioned in Chapter 1) that is similar in structure and dimensions to the cell membrane. It has been suggested that the E.R. membranes might represent complex infoldings of the cell membrane and that the cavities between the folds of the membranes may be in direct continuity with the exterior of the cell. We know that some infoldings occur, they have been mentioned before and are known as the caveolae intracellulares. These structures have a space of constant size between the two folds of cell membrane that form their walls— this space is 200 Å in width, thus the total width of the system including the two membranes of 80-Å thickness each is about 360 Å (according to Porter). Now, in some parts of the cell the membranes and spaces of the endoplasmic reticulum approach this dimension, but there is tremendous variation in the space (known as the "cisternal space") between the membranes, in parts it may be greatly dilated. This suggests that the endoplasmic reticulum is not necessarily part of the same system to which the caveolae intracellulares belong. It is probable that there is a connection between the E.R. membrane systems and the cell membrane, and Porter and other authors have recognized that the connection may not be permanent but that it may be spasmodic. A distinct possibility occurs also that the endoplasmic reticulum developed originally from infoldings of the cell membrane, that it then increased in complexity, and finally lost its connection with its point or points of origin, but temporary direct connection between the cisternae of the endoplasmic reticulum and the exterior of the cell can occur. (See Fig. 31 for summary of cell structure.)

It is thus possible that the cavity between the membranes in the E.R. represents a path by means of which the contents of the membranes or sacs (cisternae) can be excreted to the exterior.

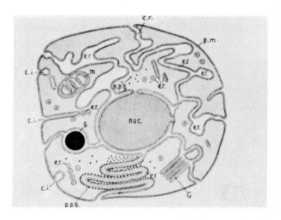

FIG. 31 Diagram of a hypothetic cell according to Robertson. He says: "The cell membrane is shown as a pair of dense lines separated by a light interzone. The invaginations of the cell surface known as caveolae intracellulares (c.i.) are indicated in several areas. Some of these extend for a considerable distance into the cell and they may connect with the endoplasmic reticulum (e.r.). The nuclear membrane is composed of flattened sacs of the endoplasmic reticulum, and by means of the nuclear pores nucleoplasm (nuc.) is in continuity with cytoplasm. The Golgi apparatus (G) is here shown as a modified component of the endoplasmic reticulum. Secretion granules (g.) are shown as dense aggregates contained within membranes of the endoplasmic reticulum. Nucleo-protein granules (n.p.g.) are shown scattered through the cytoplasm and in some regions attached to the cytoplasmic surfaces of membranes of the endoplasmic reticulum . . . One mitochondrion (m.) is shown with its cristae formed by invagination of its inner membrane." (From Robertson, *in* "Structure and Function of Sub-Cellular Components," *Biochem. Soc. Symp.,* Cambridge Univ. Press, 1959.)

In other words the secretory products of the cell pass from the cytoplasm through the walls possibly in molecular form, collect into droplets in the interior of the membranes, and pass along them. It is, of course, possible that secretion products form in the cisternae (see Fig. 32). In 1957, Hendler and colleagues found that, in the gland cells that produce albumen in the hen oviduct, the cavities or spaces between the membranes were dilated and contained a precipitate that possessed the staining and other characters of the secreted material found in the lumen of the gland. There is other evidence to indicate too that, possibly, these spaces between the membranes represent a pathway for secretory material. It is of interest, as shown by Brandes that, in the ventral lobe of the prostate gland in mice and rats, the spaces between the pairs of membranes are so enormously dilated that the membranes from opposite pairs come into close apposition with each other; pairs of membranes still

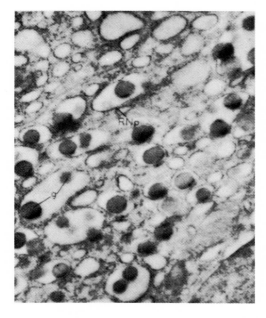

FIG. 32 Intracisternal granules in the basal region of an acinar cell of guinea pig pancreas; g, intracisternal granules inside vesicles of the E.R. These vesicles are bounded by membranes having attached ribonucleoprotein particles (RNp). (Illustration and legend from Palade and Siekewitz, *J. Biophys. Biochem. Cytol.* **2**, 671, 1956.)

exist but they are the halves of opposite pairs that have come together. The space between the original pairs of membranes is packed with an amorphous material. The ribonucleic acid granules in such pairs of membranes are, of course, on the inside instead of the outside as they were when the original halves of the pairs were together. It is possible that this enormous accumulation of material in these endoplasmic sacs really represent accumulation of secretion. It is of interest that, if the animals are castrated, the pairs of membranes with the granules on the inside separate from each other, the amount of secretory material appears to decrease, and the original halves of the membrane pairs come close together again so that the ribonucleic acid granules can now be seen on the outside. Brandes has also demonstrated that, by injection or implantation of male sex hormone, the original state of the cell could be produced again. This is in a sense an "Alice Through the Looking-Glass" type of cell in which the normal arrangement of cytoplasm and endoplasmic reticulum spaces is reversed. When this cell is in its normal condition, the spaces between the membranes contain, in fact, the strands of cytoplasm and by far the main body of the cell is occupied by what are presumably the secretion products.

It has probably not been made clear up to date that all these endoplasmic sacs are probably continuous with each other and thus form a convoluted system of spaces coursing through the cytoplasm of the cell. It appears, also, that folds of the endoplasmic reticulum form the nuclear membrane (this will be discussed later). Since this membrane has been shown to contain pores, it is quite possible that material could pass from the nucleus into the cisternae and along these channels to the exterior of the cell without having to traverse the cytoplasm or the cell membrane at all. A possible origin of the endoplasmic reticulum from the nucleus is shown by Fig. 33.

At this point we should enlarge a little on the subject of the caveolae intracellulares. These were originally described by Yamada as consisting of a number of infoldings of the surface membrane of some cells. He found them in endoneural, endothelial, pulmonary epithelial, and muscle cells. Although it is possible that these caveolae could, in fact, represent the regions where the E.R. channels communicate with the exterior, the existing evidence is against this. If this were so then there is a continuous, if extremely tortuous, pathway that stems from the vicinity of the nuclear membrane ex-

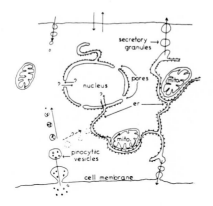

FIG. 33 Principal organelles in a cell according to Siekewitz. (From "Regulation of Cell Metabolism," *Ciba Foundation Symp.*, Churchill, London, 1959.)

tending to the exterior of the cell that can be described as a possible circulatory system for the cell. At the moment it seems most likely that the caveolae represent only blind pockets that extend for variable distances into the cells, and there is no evidence that they are actually continuous with the ergastoplasm. Palade has suggested that these invaginations in endothelial cells, at any rate, represent stages in a process of what is described as "micropinocytosis," which has already been discussed. It has been suggested that when the little vesicles (formed by the caveolae) containing water pass into the cell, the cell membrane then dissolves away and the water is liberated into the cytoplasm. This conception of the cell membrane "pinching off" and a little vesicle passing into the cytoplasm is attractive, but one of the difficulties of accepting it is the mechanism of the formation of the vesicle. We know that the cell membrane can invaginate in this way (it has been seen to do so) but whether, when the membranes on the two sides of the vesicle bend around and come in contact with each other, they are, in fact, capable of coalescing and nipping off a vesicle is another matter. It has been pointed out that, if the membrane of a cell were a purely lipid substance, there would be no difficulty at all in such a concept, but, since the outside and probably the inside parts of the lipid membrane of a cell are covered with protein, this would make such a coalescence rather difficult to conceive from a physiochemical point of view. On the other hand, an invagination of the membrane may draw off a little droplet of water and the membrane may then burst internally and squirt the drop of water into the cell, and then there would be no difficulty in the burst sides of the membrane joining up in the usual way.

The Golgi apparatus has also been shown to be composed of double membrane structures without, however, the associated ribonucleoprotein granules of the endoplasmic reticulum. Although we will be dealing with the Golgi apparatus later on, we might mention here the concept that has been put forward by some authors that the Golgi apparatus represents a part of the system of the endoplasmic reticulum and that it is in communication with it. We do not know whether the whole of the endoplasmic reticulum is in continuous communication with itself, however, if it is and if the Golgi apparatus is part of it, then it is possible to assume that everything passing along the cisternae of the reticulum will have to filter through the Golgi region. It should be stressed that there is no proof that this is so, that many cytologists believe it is not so, but that such a possibility exists.

One problem that should be discussed at this point is the origin of the endoplasmic reticulum. Where in fact does it come from? There are many controversial theories on this subject. It has been suggested by Palade that it might originate as an infolding or infoldings of the cell membrane.

The Nebenkern, because of its high concentration of E.R. membranes, might be considered as a possible region in the cytoplasm where this material is being formed. The Nebenkern is made up of a dense aggregation of concentric rings, and one could, perhaps, consider the possibility of the membranes peeling off from such a germ center. The same sort of appearance has also been noted in association with the nuclear membrane and has led to the suggestion that the endoplasmic reticulum is formed in this region and that the layers are, in fact, peeling off the nuclear membrane. However it may quite easily be interpreted the opposite way, i.e., the close apposition of the existing ergastoplasm around the nuclear membrane may be a part of the complex canalicular system that permits an almost direct passage through the nuclear membrane direct into the cisternae of the reticulum. Again, there are a number of authors who think the endoplasmic reticulum is associated with mitochondria. There is no doubt that many pictures of mitochondria, closely surrounded by concentric layers of endoplasmic reticulum have been obtained, and we, ourselves, have found this particularly well demonstrated around the mitochondria of the liver cells of scorbutic guinea pigs. In some cases, Sheridan has found a tremendous concentration of many layers of reticulum around the mitochondria, as though, in fact, the reticulum is being formed on the surface of the mitochondrial membrane and is being split off. Rouiller and his colleagues have demonstrated that, in animals which have been poisoned, the endoplasmic reticulum, destroyed or badly damaged by the treatment, always reappears in association with mitochondria. The membranes of the reticulum may not undergo direct physical formation in the sense that the membranes are produced on the surface of the mitochondrial membrane and split off; it may be that the mitochondria supply the energy necessary for the production of these membranes. Furthermore the close relationship between the endoplasmic reticulum and the mitochondria may simply be physiological, that is they are cooperating in some metabolic process such as the synthesis of protein. However, the origin of the endoplasmic reticulum is a problem that is far from solved, and we must await further evidence before drawing a conclusion.

FOUR
Mitochondria

N the latter half of the 19th century, cytologists described a number of structures in cells that were probably mitochondria. Kolliker in 1850 described structures in muscle that were probably sarcosomes (muscle mitochondria). In the early part of the decade 1880–1890, Flemming described in cells some structures of a filamentous nature that were probably mitochondria, their official discovery is, however, usually attributed to Altmann in 1886, and they were put more or less definitely on the cytological map by Benda in 1903. They can be easily displayed with suitable dyes. Altmann's aniline fuchsine–picric acid technique shows them up very well. Regaud's iron–hematoxylin method is also very good. The mitochondria can even be seen in the living cell by staining them intravitally with Janus green and, of course, they show up extremely well with the phase-contrast and interference microscopes. There is not much difficulty, then, in establishing their existence and nature. Their dimensions vary, of course, but they range in most cells from about 0.5 to 2 μ or longer. The filamentous mitochondria, for example, which are present in connective tissue cells, may be much longer than this. Extremely small, and even submicroscopic mitochondria may also exist. Extremely small-sized mitochondria are, in fact, known, and Rhodin (1954) has described structures ranging from 0.1 to 0.5 μ that are presumably mitochondrial in nature. Green has pointed out that the various enzymes and proteins that mitochondria contain can be accommodated in a very much smaller particle than any known mitochondria and still perform all the functions required of them. Some of these much smaller mitochondria may not have all the details of organization of larger mitochondria, but there is no reason why they should not contain all the enzymes and other compounds that are important for mitochondrial structure and function.

There have been some suggestions that minute mitochondria are present in microorganisms. There is no doubt that they are present in yeasts, fungi, algae, and protozoa; but it seems unlikely that they exist in bacteria since, although very small mitochondria are known, they are still not much smaller than the entire bacterium. Many years ago it was in fact suggested that mitochondria really represented a sort of symbiotic bacteria. The original promoters of this hypothesis called down on their heads some unkind remarks, but the unique enzyme independence of mitochondria that has been demonstrated in recent years causes one to think twice about this suggestion, and the article by S. Nass recently published in the *International Review of Cytology* (1968) and entitled "The significance of the structural and functional similarities of bacteria and mitochondria," raises the whole matter again in a more acute fashion. He points out that the mitochondrion is "the only structure in the nucleated animal cell which satisfies both the biochemical structural criteria for an evolutionary relationship with bacteria." He goes on to say that mitochondria have most of the requirements for an independent existence, although they are not able to synthesise some essential compounds, but he points out that many symbiotic and parasitic organisms have lost functions they obviously possessed prior to evolving their dependent form of life. He believes that the evidence available does in fact suggest that the mitochondria may have an evolutionary relationship to bacteria.

In the cells of animals and plants the distribution of mitochondria varies with the function of the cell, for example, in epithelial cells mitochondria are often oriented with their long axes parallel with the long axis of the cell. In pancreatic cells, mitochondria may encircle a lipoid droplet, they may be distributed more or less evenly through the cells or even aggregated mainly in the base or in the apex of a cell. Lehninger points out that they may also be found near structures that need the ATP (adenosine triphosphate)

that they produce (e.g., in muscle) where they are in close association with the myofibrils, or in spermatozoa where they have a helical
arrangement around the middle piece and one thus ideally situated
to pass ATP to the tail, which is the contractile organ of the sperm,
they are also found near the bases of the cilia in ciliated cells.

In cell division, mitochondria tend tto break up into small granules during the prophase and become passively distributed more
or less evenly between the two daughter cells.

It is of interest to note that, in living tissue culture cells,
the mitochondria undergo continuous movement. This has been noted
by quite a number of workers, and the movements have been described frequently. These movements may be fitful individual movements by the mitochondrion itself, i.e., bending, wriggling, and with-
ing, or may consist of movement of the whole mitochondrion through
the cytoplasm of the cell. The cause of the bending and twisting
movements of the mitochondria is not exactly known. At one time
it was thought to result from the contracting of the polypeptide
chains that are part of the membrane of the mitochondria. It may
be due to the fact that the mitochondria are engaged in active
ion pumping or being twisted by cytoplasmic movements. In the second type of movement in which there is a transport of the whole
mitochondrion from one part of the cell to another, the movement
is possibly related to streaming movements of the cytoplasm or
to electrical forces. In tissue culture cells the mitochondria have been
described as making journeys from the cell membrane to the nuclear
membrane and back again almost as if they were discharging either
an electric charge or even perhaps some compound on the membrane
that they touch. There are also records that, when the mitochondria
come in contact with the nuclear membrane, the nucleolus sometimes
moves across the cell and comes in contact with the nuclear membrane at the same point at which the mitochondrion is touching
it, and this suggests that there may be some exchange of compounds
or substances during this period. This has been demonstrated in
spinal ganglion cells by Dr. Tewari in the author's laboratory (see
Figs. 34A, B and C). Filamentous mitochondria have frequently been
observed to break up into batonettes, and these have been observed
to break up into granules. The reverse process has also been found
to occur, granules have been seen to join up into batonettes and
these into longer filamentous mitochondria. What the significance
of this breaking up is, we do not know, but a little later we will

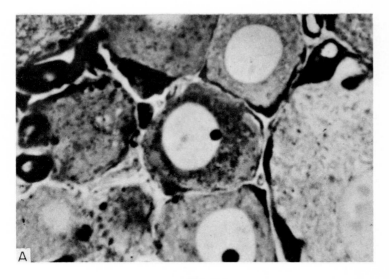

FIG. 34A

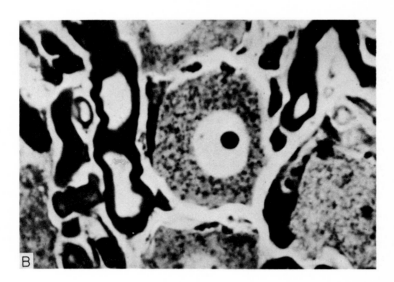

FIG. 34B

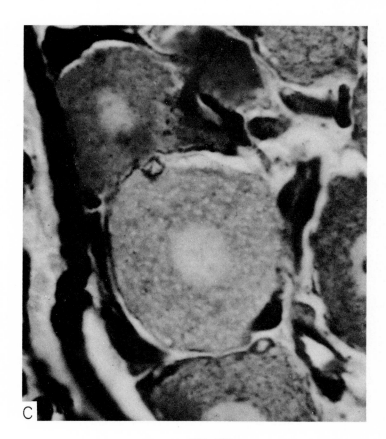

FIG. 34C

FIG. 34 Relation between nucleolus and mitochondria in spinal ganglion neurons. (A) Nucleolus in contact with nuclear membrane and mitochondria congregated around nucleus especially on the side where the nucleolus is touching, (B) nucleolus in center of nucleus and mitochondria scattered through cytoplasm, (C) decrease in number of mitochondria and presence of many vacuoles of unidentified material, faint diffuse stain by nucleus. It is of interest that the nucleolus reacts with the same intensity as the mitochondria to the Regaud technique. Presumably this indicates that it contains a similar lipoprotein complex. (Preparation and photograph by H. B. Tewari, Dept. of Anatomy, Emory Univ.)

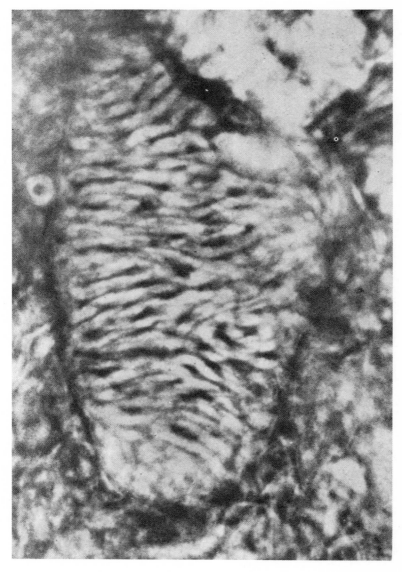

FIG. 35 Intact whole mitochondrion from heart muscle, unfixed, in a drop preparation and negatively stained by sodium phosphotungstate. The cristae can be seen in this preparation. (From F. S. Sjöstrand, Ultrastructure and Function of Cell Membranes, *in* "The Membranes," Academic Press, 1968. Magnification 120,000×.)

refer to this process again in the light of what will be said about the metabolic activities of mitochondria.

When mitochondria were studied in ultrathin sections of tissue, it was found by Palade in 1953, Sjöstrand, 1953, and by Rhodin in 1954, that they had an interesting internal structure (see Fig. 35). These workers found first of all that there was a single membrane round the outside, that inside this membrane was another membrane so that the two made a double membrane and, after a little controversy, it was agreed that the inner of these two membranes was extended to form a number of bars or plates that projected into the interior of the mitochondria (see Fig. 35), in some cases touching or almost touching the other side. These plates or bars or tubes had double structure again, since they were a reflection of the inner membrane of the mitochondria. They were described by Palade as the "cristae mitochondriales" (see Fig. 36). Changes in these cristae appear to be related to function; for instance, Palade has drawn attention to the fact that the amount of cytochrome in the mitochondrial fraction of a homogenate is directly related to the number of cristae present in the mitochondria. When metabolic activity is high, for example, in rapidly contracting skeletal

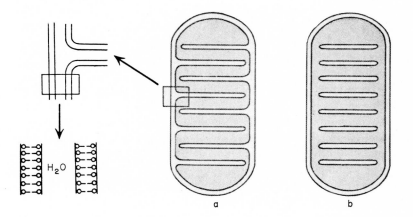

a   b

FIG. 36 Structure of mitochondria: (a, b) the pattern of organization of mitochondria; (a) viewpoint of Palade, (b) that of Sjöstrand. The rectangle indicated in (a) is enlarged to show details of membrane structure. The probable molecular structure of each of these membranes is indicated below. The gap between the membranes is probably highly hydrated. (Legend and figure from Lehninger, *Sci. Am.* **202**, No. 5, 102, 1960.)

and heart muscle, the mitochondria have many densely packed cristae. In smooth muscle, however, where the activity is greatly reduced, the cristae present in the mitochondria are relatively sparse. There are claims that, e.g., in mouse tumor cells and in paramecium, cristae may be everted into the cytoplasm. Presumably in this case the outer membrane is folded into cristae. See Fig. 40 for the appearance of mitochondria in adrenal cortical cells.

In most electron-microscope pictures of mitochondria, there appears to be a space between the outer and inner membranes of the organelle; but there is a possibility that this space is an artifact and that the inner and outer membranes are in fact fused together. Pancreatic cells fixed in permanganate and embedded in Vestopal show the two membranes in the latter state.

The cristae may extend all the way across the mitochondrium, fusing with the membranes of the other side (septate cristae), or may extend to varying degrees towards the other side (complete cristae) or they may be broken up with holes or notches (incomplete cristae)—sometimes the cristae are platelike in nature as, for example, in the flight muscle of the dragonfly where the plates may be as much as 8 μ in length. In some muscles the mitochondria (sarcosomes) have been seen to extend completely around the "I" bands of the myofibrils in the form of a ring. The villous type of cristae extend into the interior of the mitochondrion like fingers that have a roughly circular shape; however, in the mitochondrion of the organs of some animals (e.g., cricothyroid muscle of the bat) they are triangular in cross section.

The presence of cristae in mitochondria is obviously a device for increasing the surface area. This is important since the membranes of the mitochondria including those of the cristae carry the respiratory enzymes of the cell. Lehninger has made a series of calculations to demonstrate the significance of the cristae in increasing the total surface area of the mitochondria. He calculates that the surface area of an average liver cell is about 3000 μ^2. A single liver cell contains about 1000 mitochondria, which are predominantly circular in shape and about 1.0 μ in diameter. Each of these mitochondria can be shown to have a surface area of about 13 μ^2 so the total surface area of mitochondria in each of these cells is about 13,000 μ^2. Assuming that each mitochondrion has 10 cristae, each of which can be calculated to have an area of 16 μ^2, the total surface area of the cristae of the mitochondria of one cell is 16,000 μ^2. Add-

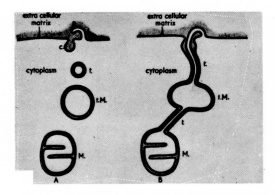

FIG. 37 Possible origin of mitochondria by infolding of cell membrane according to Robertson. (*In* "Structure and Function of Sub-Cellular Components," *Biochem. Soc. Symp.,* Cambridge Univ. Press, 1959.)

ing this to the surface area of the mitochondria themselves, we get a total mitochondrial surface for the cell of 29,000 μ^2, which is nearly ten times the total surface area of the cell. The liver-cell mitochondria are not particularly rich in cristae so that one requires very little imagination to conceive of the surface area of the mitochondria of some cells where each mitochondrion may have 20, 30, or 40 cristae.

It is of interest that the double membranes that surround the mitochondria show a similar 80-Å unit structure that is characteristic of the cell membrane, and Robertson has suggested that it is possible that the mitochondria may have originated by an invagination of the membrane that was nipped off to form the mitochondrion (see Fig. 37). This is an interesting suggestion but it is not readily acceptable to most cytologists. The membranes of the mitochondria appear to consist of two protein layers separated by a double layer of lipids. This is again characteristic of cell membrane structure except that, unlike the cell membrane, it is symmetrical.

The chamber enclosed by the mitochondrial membranes contains fine granules in most electron-microscope preparations, but it may contain large granules or even crystals. The "matrix," as it is called, is thought to be semisolid; and in fact there is some evidence that it contains DNA (deoxyribonucleic acid) (see below). There is also evidence that the granules of the matrix have the ability to bind magnesium and calcium ions.

Bensley in 1934 was the first one to isolate mitochondria by differential centrifugation, and he was able to demonstrate that mitochondria contained a considerable amount of protein, lipid, and fat. Bensley's chemical composition of mitochondria is

Proteins and unknowns	65%	Lecithin and cephalin	4%
Glycerides	29%	Cholesterol	2%

Subsequent studies have shown up to 30% of phospholipid (lecithin and cephalin). Mitochondria undergo a very considerable change under various conditions. Bensley claimed that mitochondria (as demonstrated by standard staining reactions) were greatly reduced in liver cells of starved animals, but Sheridan in our laboratory and Fawcett have demonstrated by electron-microscope techniques a great increase of mitochondria in starvation. Sheridan has further demonstrated a still further increase of mitochondria in the liver cells of guinea pigs suffering from scurvy. Mitochondria are also affected in number and form by narcotics, enzyme poisons, old age, and so on. Sometimes mitochondria store unusual compounds, and sometimes they store compounds that are usual, but to an abnormal degree. The double membrane enclosing the mitochondria is of a semi-impermeable nature and, as a result of this, mitochondria have been described as osmometers. There is no doubt that molecules of various sizes pass with different degrees of readiness through the mitochondrial membrane. It has been shown that large molecules pass through fairly easily while some small molecules pass through either not at all or only with very great difficulty. It has even been said by Emmelot and Bos (1956) that this permeability, particularly of liver mitochondrial membranes, is affected by thyroxine. Also there are claims by some authors that mitochondrial membranes have pores. Mitochondria may accumulate water and swell. They may do this in starvation and in various pathological conditions. When this occurs they produce the condition that the pathologists describe as cloudy swelling. Cloudy swelling has been known to occur sometimes in neoplasms and sometimes after poisoning with various toxins. The mechanics of cloudy swelling are that the mitochondria become very large and lose their cristae, the matrix stains less strongly, and sometimes the outer membranes of the mitochondria may disappear and the mitochondria may fuse and so form structures known as chondriospheres. Under some pathological conditions mitochondria may become disrupted, undergo lysis, and portions may be ejected from the cell. Mitochondria of *Paramecia* can be seen in Fig. 38.

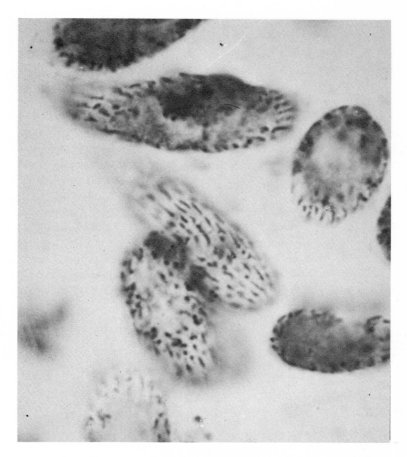

FIG. 38 Mitochondria in *Paramecium*. (Photograph by K. C. Richardson.)

Mitochondria are capable of abnormal storage. Accumulation of ferritin by mitochondria was demonstrated by Kuff and Dalton in 1957, and they may also accumulate iron pigments or bile pigments. Brachet has described ferritin particles in young erythroblasts where he says they penetrate the mitochondria and then break into smaller iron-containing particles. When the mitochondria eventually burst, the granules become scattered throughout the cytoplasm and take part in the formation of hemoglobin. Silver granules and keratinous materials have been also seen inside mitochondria. They have been found to contain melanin, and Graffi, in a series of papers from 1939 to 1941, claims that they also contain or store carcinogenic hydrocarbons. Sometimes this storage affects the function of the mitochondria, and sometimes the accumulation of material is due to malfunction of the organelle. They may also store neutral fats or lipids. In plants mitochondria have been recorded as storing starch. As a result of storage mitochondria undergo physical changes. The membrane may become single and the cristae decrease in size and may be lost altogether. If the stored substance is eliminated, the mitochondria resume their normal appearance. There is a vast literature on the the role of mitochondria in the production of almost any kind of product—protein, yolk, fat, glycogen, and so on—but this is not the place to review this work. For further information see Bourne, Mitochondria and Golgi Complex in "Cytology and Cell Physiology," 2nd ed., Oxford Univ. Press, 1951; 3rd ed., Academic Press, 1963. See Fig. 38 for mitochondria in *Paramecium.*

In the past there were many scornful remarks in the literature about the supposed synthetic activity of mitochondria, but the study of the enzymic equipment of mitochondria shows that they have

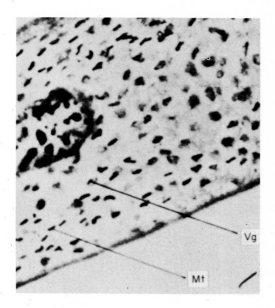

FIG. 39 Apparent synthetic activity by mitochondria (Mt) vegetative granules in the protozoan *Opalina.* It is suggested by Horning and Richardson that the material produced is protein. (*Arch. Exptl. Zellforsch.* **10,** 488, 1929.)

the equipment to effect a wide range of synthetic activities (see Fig. 39).

It has been mentioned that the isolation of mitochondria by the differential centrifugation of homogenized cells was first carried out in 1935 by Bensley and by Hoerr. It was not until 10 or 15 years later that this technique became used extensively and this time by biochemists who set to work to find out the metabolic composition of mitochondria, i.e., the nature and content of enzymes concerned with metabolic processes.

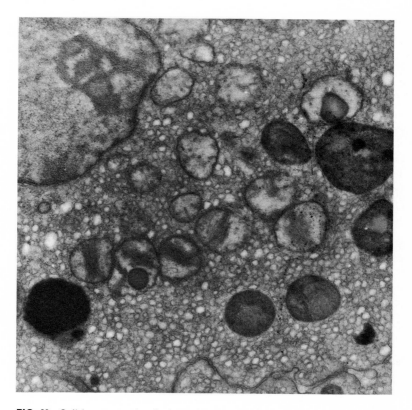

FIG. 40 Cell from zona fasciculata of guinea-pig adrenal cortex. Many typical mitochondria are seen. There are also a number of bodies containing small spherical multilaminate myelinlike bodies. Are these aberrant or modified mitochondria? Note many mitochondria show imperfections in their membranes. (Preparation and photograph by R. Q. Cox, Dept. of Anatomy, Emory Univ.)

THE CHEMICAL NATURE OF MITOCHONDRIA

Originally, it was suspected that mitochondria contained a good deal of lipid material. One finding that suggested this was that mitochondria were stained by a method that is similar to that used for staining *Mycobacterium tuberculosis* and *Mycobacterium leprae.* These bacteria are stained by treating them for some minutes with a hot phenolic solution of basic fuchsine and once stained in this way they resist the destaining effects of acid alcohol. This is the origin of their designation as acid-fast bacteria, and their staining idiosyncrasies are believed to be due to a waxy or lipoidal coat. Mitochondria stained by hot acid fuchsine resist the decoloring action of picric acid for longer periods than most other cell constituents, and this suggests that the mitochondria might also have a membrane that contains material of a lipoidal nature or at any rate is of composition similar to that of the bacteria mentioned.

Mitochondria are difficult to demonstrate with conventional methods of staining if the fixative has contained acetic acid, alcohol, ether, chloroform, acetone, or other fat solvents. Also mitochondria have been found to stain with osmic acid, and, incidentally, to resist destaining by extraction with turpentine—a property of certain types of lipoprotein complexes. Baker has found that his "acid-haematin" test for lipids applied to a variety of tissues, nearly always gave a positive reaction with the mitochondria. Thus there had accumulated for some years direct and indirect evidence that mitochondria contained appreciable amounts of lipid. Bensley's experiments, which, incidentally, were carried out some years before Baker's staining studies, showed that mitochondria that had been produced by homogenization and differential centrifugation contained appreciable amounts of lipid (30%).

METABOLIC SUBSTANCES FOUND IN MITOCHONDRIA

Perhaps we should give a brief summary of what was suspected about the metabolic significance of these organelles prior to the revolutionary studies that were made with isolated mitochondria beginning during the middle of the 1940's. One of the characteristic staining reactions of these organelles is that they give a green or green-blue stain with Janus green B, which is diethylsafranine azodimethylaniline, and it has been claimed by Cowdry that this reaction is primarily due to the diethylsafranine monocarboxylic acid component since this compound gives a very good and specific reaction with these structures. It was observed many years ago by T. B. Robertson that if one drop of a saturated solution of safranine was added to a solution of trypsin, a colored precipitate was formed, and subsequently it was demonstrated that this colored precipitate had proteolytic activity. Then Marston demonstrated that other azo dyestuffs would react in a similar way, particularly neutral red which is a dimethyldiaminotoluazine hydrochloride. Marston suggested that these results indicated that the reaction of mitochondria with Janus green might signify that they contained proteolytic enzymes. However, the green stain does not persist and eventually the organelle turns pink. The concept of the staining reaction with Janus green has now been pretty well proved by Lazarow and Cooperstein to indicate that mitochondria play an important part in cellular oxidations and that the production of a pink color from the Janus green by the mitochondria is due to the DPN specific dehydrogenases. The fact that Janus green B does not stain the mitochondria permanently green but that the green color gradually becomes reduced to the pink and then to the colorless form has been known for many years.

Joyet-Lavergne demonstrated more than twenty years ago that mitochondria contain an oxidase system which oxidizes cobaltous to cobaltic salts and that these latter stain the mitochondria green. It was noted by Gatenby that in the snail, *Lymnaea,* the mitochondria are colored yellow in the natural state, presumably by a carotenoid pigment; also, extracted mitochondria from the organs of some animals frequently have a yellow appearance that is probably also due to carotene. It is of interest in this connection that it has been demonstrated by the present author and by Joyet-Lavergne that mitochondria give a blue reaction with antimony trichloride in chloroform solution, an acknowledged reaction for vitamin A. Carotene is also a provitamin A, so these histochemical results indicate the presence of vitamin A in mitochondria. Criticism of Joyet-Lavergne's results was made by Gomori because the former author had used alcohol as a fixative. The present author has, however, always applied antimony trichloride in chloroform solution direct to fresh unfixed tissues and this suggests that this result is a true one; in any case, biochemical tests have now confirmed (see Goerner) that mitochondria (prepared by homogenization and differential contrifugation of the cells) contain 27–32% of their weight of lipids and that 100 mg of this lipid contains approximately 249–910 USP units of vitamin A. Joyet-Lavergne has suggested that mitochondria contain a redox system in which vitamin A plays a part.

A number of authors—Leblond, the present author, and Giroud and his co-workers—have demonstrated that mitochondria of some organs react with acetic acid–silver nitrate solution (which has been demonstrated as being a specific reagent for vitamin C) to give a black reaction, and this indicates that vitamin C may be present in them. One has to accept the intracellular localization of vitamin C with a certain amount of discretion in view of the very destructive effect of this reagent on the cell cytoplasm. Electron-microscope studies in the present author's laboratory have demonstrated that this reagent has a most drastic effect on the ultrastructure of the cell, and we cannot be sure that the localization of vitamin C in the mitochondria is, in fact, a real thing. It is of interest, however, that Chayen has found that the mitochondria of plant cells give this reaction very intensely and very specifically. It may be that mitochondria in plant cells and some animal cells do, in fact, contain vitamin C, but we need further studies before this can be confirmed. Other vitamins appear to be present in mitochondria, members of the vitamin B complex have been found to be present, in some cases, in the form of coenzymes. The actual vitamins of the B complex recorded are vitamin B_1 (thiamine), riboflavin, nicotinic acid (niacin), pantothenic acid, and pyridoxine. However, although these vitamins are present in mitochondria, they are not present in any greater concentration than in the other parts of the cell so that they are not exclusively contained in these organelles. These studies were made on cell homogenates and, of course, it is possible that the presence of the vitamins in the other fractions of the homogenate may be due to the fact that they have been leached out of the mitochondria by the solution that is used in the homogenization process in these particular experiments, and further work would have to be done before we could be certain about this.

The present author, Joyet-Lavergne, and Giroud have demonstrated that mitochondria contain glutathione or protein-bound SH, and this is further evidence that mitochondria can play an important part in oxidation–reduction mechanisms in the cell. It is of interest that mitochondria have been found in large quantities in the phloem cells of plants, which are concerned with transport and may be concerned with this process. This suggestion is made in view of

Conway's views that redox systems can play an important part in ion pumping. Another substance which was demonstrated histochemically in mitochondria by the present author, using the Schultz reaction, was cholesterol. This was particularly evident in the mitochondria of cells of the adrenal cortex. Similar but less intense reaction was shown by the mitochondria of the liver, and it has been claimed that isolated liver mitochondria contain about 2% cholesterol, but not everyone agrees with this.

Bensley has recorded the presence of a red pigment in the mitochondria and also in the submicroscopic particles of the liver cell. He believed that this pigment is derived from the oxidation of unsaturated fats and possibly the phospholipids of the mitochondria, and this led him to suggest that in the liver cell the mitochondria in particular are, possibly, the seat of highly active oxidative processes that involve the metabolism of fats. The possible relation of mitochondria to oxidation—reduction processes was indicated by the publications of Ludford. He demonstrated that, if methylene blue was added to tissue cultures, the mitochondria of the cells stained a brilliant blue color but this could be inhibited by potassium cyanide. If the cells were exposed to a bright light, the blue color was rapidly bleached.

It is of interest that the mitochondria, although they have been demonstrated to contain a good deal of fat and lipid, do not give a positive reaction with Sudan III. The membranes of mitochondria have been found to contain three varieties of phospholipid. These are: cardiolipin, phosphatidyl ethanolamine, and lecithin. It is of interest that the latter two compounds are found in the membranes of the endoplasmic reticulum, cardiolipin is not.

Mitochondria also contain protein (quite a high proportion of it), but it is of interest that the earlier workers, using Millon's reagent, produced a negative reaction for protein. Subsequently Bensley and Gersh using a Millon's reagent of a different formula were able to obtain positive results from the mitochondria of many tissues, and they found that they were particularly well stained in frozen, dried sections and particularly in those of *Amblystoma* liver. The same authors demonstrated that, in undenatured, frozen, dried sections of *Amblystoma* liver, the mitochondria were destroyed if the sections were exposed to artificial gastric juice and artificial pancreatic juice.

Despite all this chemical information, most of which was in

existence by the beginning of the 1940's, there was no certainty as to the function of mitochondria, and it was not until 1946 and 1947 that the late George Hogeboom and his colleagues at Bethesda established on a sound basis our ideas of their function by demonstrating that the major proportion of the succinic dehydrogenase and an appreciable proportion of the cytochrome oxidase activity of the liver cell were present in the mitochondria. This at once suggested that these organelles were the major sites of aerobic respiration in the cell. It is of interest that as long ago as 1915, Dr. Kingsbury had suggested that mitochondria were concerned with this process. His reasons for this were largely due to his observations that anesthetics such as ether and chloroform, which depressed cellular respiration and the respiration of the animal in general, also broke up mitochondria in the cell.

The work of Hogeboom and his colleagues was carried out on homogenates of liver that had been produced by grinding up liver with saline. This gave poor preservation of the form of the mitochondria and subsequently 0.8 *M* sucrose was used—this preserved the nature and form of the mitochondria very well. In the beginning there was some doubt as to whether the material being assayed was in fact mitochondria or not, but when sucrose was used this doubt gradually disappeared. Eventually electron-microscope studies of the structure of the granules isolated by homogenization and differential centrifugation demonstrated beyond doubt that they contained the same structures as the mitochondria of normal cells. Another possible source of error, however, began to haunt the biochemists and was frequently verbalized by cytologists and this was that the mitochondria did not really contain the oxidative enzymes mentioned above, but that they were being absorbed or adsorbed by them from the homogenate. This possibility was negated at least in part by adding more enzyme to the homogenate and demonstrating that it could be recovered almost 100% from the supernatant and so was not taken up by the mitochondria. Thus the localization of the enzymes in the original cell was probably in the mitochondria. Subsequently, Kennedy and Lehninger demonstrated that in addition to containing cytochrome oxidase and succinic dehydrogenase, mitochondria also catalyze the condensation of acetate with oxaloacetic acid, the formation of succinic acid from α-ketoglutaric acid, and the formation of malate from succinate. These three reactions represent three steps of fundamental importance in the Krebs tricarboxylic acid cycle of which we will have more details shortly. Mitochondria also contain coenzyme I (diphosphopyridine nucleotide, DPN, or NAD) of which nicotinamide is an important constituent and cytochrome reductase (flavin adenine dinucleotide or FAD), which is a flavoprotein. These two enzymes are links between the Krebs tricarboxylic acid cycle and the cytochrome system, and it, therefore, appeared that mitochondria probably contained the whole enzymic equipment necessary for aerobic respiration of the cell. In fact, it was originally demonstrated that a centrifugate of cells that included only nuclei and mitochondria were capable of carrying through the whole of the oxidation of glucose to CO_2 and water; subsequently it was demonstrated that isolated mitochondria on their own could do this (see Fig. 44). Transaminase activity was subsequently demonstrated in mitochondria, although in only the same concentration as in the rest of the cytoplasm. Transaminase is concerned with protein synthesis and it is of interest that pyruvic, oxaloacetic, and α-ketoglutaric acids are the corresponding α-keto acids of the amino acids, alanine, aspartic and glutamic and need only to be transaminated to produce them. Since the former compounds are themselves formed as intermediates in the course of the Krebs cycle, this indicates a mechanism whereby mitochondria might synthesize fresh protein material and so increase in size themselves or synthesize protein for other parts of the cell. This possibility is further extended by the fact that both DNA and RNA are known to be present in varying amounts in mitochondria. There is no doubt that mitochondria do have the equipment for synthesizing protein, although to what extent they do this *in vivo* is not known. It has been found that both liver and heart mitochondria (and probably also those of other organs) can incorporate labeled amino acids into the proteins of mitochondrial membranes especially into a fraction which contains phospholipid, RNA and succinic dehydrogenase, they were not found in the cytochrome C or malic dehydrogenase containing fraction. According to Chevremont DNA can be found in mitochondria at certain stages of cell division. According to Lehninger (1965) "each membranous structure of the cell including the mitochondrion may have the enzymatic capacity to 'spin off' new membrane in such a manner that pre-existing membrane serves as a template for its own assembly from specific proteins and lipids. The enzyme molecules may be synthesized by some more

specific, possibly ribosomal, process and later attached to the mitochondrial membrane in a secondary process." It seems now that
DNA is regularly found in mitochondria and RNA is also present.
Dr. M. Steinert of the University of Brussels claims that, in the inner
chamber of the mitochondrion, ribosomes, transfer RNA, amino
acyl, and RNA synthetases are present. There is evidence that the
proteins synthesized by mitochondria above are primarily structural
proteins, the synthesis of other proteins such as the cytochromes,
seems to require, in addition the cooperation of extra mitochondrial
synthetic systems (C. P. Henson *et al.* at the 1968 Federation So-
cieties Meeting). It is of interest that the protein synthetic ability
of mitochondria resembles more that of bacteria and appears quite
distinct from that of the rest of the cytoplasm. The reader is referred
to the earlier references to mitochondria and bacteria. The DNA
found in mitochondria differs from normally accepted more or less
straight helical structures of DNA as demonstrated in the Watson–Crick model in that it has closed circular structure and has
no free ends. It also has additional twists like a many-twisted figure
8. Messenger RNA also appears to be passed into the cytoplasm
from the mitochondria and is different from that produced by the
nucleus and has a structure similar to the circular DNA. The presence of DNA in mitochondria provides a basis for cytoplasmic
inheritance.

All the fatty acid oxidase activity in the cell is also in the
mitochondria and so is 80% of the octanoxidase activity. Since these
discoveries, an enormous amount of work has been done on the
mitochondrial enzymes and, in addition to those described, cysteine
desulfhydrase, xanthine dehydrogenase, choline esterase, histami-
nase, aryl sulfatase, choline oxidase, betaine dehydrogenase, and
many others have been found to be present in these organelles.

It is of interest too, that a considerable amount of ribonuclease
and deoxyribonuclease are present in mitochondria.

It was demonstrated by fragmenting mitochondria and separating the membranes from the contents that cytochrome oxidase
and succinic dehydrogenase activity were located primarily in the
membranes. Subsequently it has been shown that the ability of
the mitochondria to carry out the Krebs cycle and to combine
this with oxidative phosphorylation is dependent on the presence
of intact double membranes in the mitochondria. Now, it has been
demonstrated and mentioned earlier that in the structure of mito-

FIG. 41 Reconstruction of a mitochondrion. The two membranes, external and internal and the cristae are clearly seen. The outer and inner membranes and the membranes of the cristae are covered with many thousands of spheroidal particles. ATP synthesis is believed to take place predominantly at the level of the internal particles. Cells may contain a thousand mitochondria—the gaint ameba *Chaos chaos* is said to have half a million. Mitochondria are machines of high energy production. The most modern steam engine converts approximately 8% of the fuel it burns into energy, the mitochondrian has an energy yield of 50%. (Figure and legend from *Rassegna Med.* **42,** No. 2., 1965.)

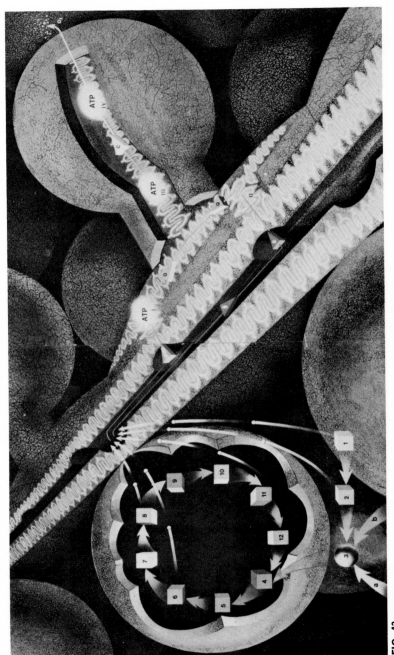

FIG. 42

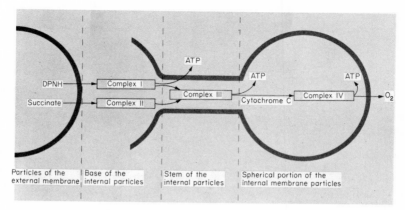

FIG. 43 Diagrammatic representation of the processes illustrated in Fig. 42.

chondria there is an outer membrane and an inner membrane to the mitochondrion, so that it is surrounded by a double membrane. Also, the inner membrane is folded inward to give elongated cristae, and these cristae, because they consist of a fold of the inner membrane, are themselves double membranes.

The presence of respiratory enzymes in mitochondria has been already described in this chapter. From their presence it is obvious that the mitochondria can metabolize proteins, fats, and carbohydrates to CO_2 and O_2 with the production of energy-producing ATP (adenosine triphosphate). There are several mechanisms responsible for this chain of events.

FIG. 42 Details of a section through portion of a mitochondrion. Note particles on inner side of membrane are stalked, those on outer side are not. The lipoprotein structure is shown in the mitochondrial membranes. The lipids are the whitish fork like areas in the interior of the membranes. The protein is the granular reticular material surrounding the lipid. The external particle is said to contain the Kreb's cycle enzymes (1. Lactic acid, 2. pyruvic acid, 3. acetyl-CoA, 4. citric acid, 5. cis-aconitic acid, 6. isocitric acid, 7. oxalosuccinic acid, 8. alpha-ketoglutaric acid, 9. succinic acid, 10. fumaric acid, 11. malic acid, 12. oxaloacetic acid, a. lipids, b. amino acids.) The Krebs cycle makes available a number of electrons, shown as small white spheres with tails, to the respiratory chain shown mainly in the stalked particles of the internal membrane. Transfer of these electrons to complexes 1 and 11 is carried out by DPNH and succinate (5) coenzyme Q (Q) figures in the respiratory chain and cytochome is indicated by "C." The location of ATP formation is shown in the figure. (Figure and legend from *Rassegna Med.* **42**, No. 2, 1965.)

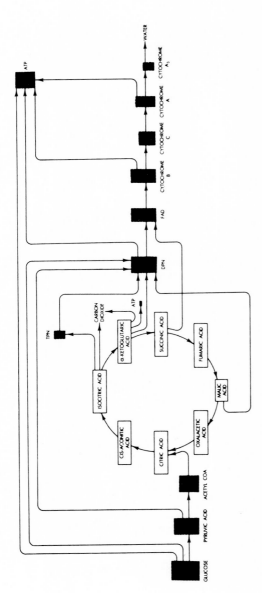

FIG. 44 The chain of oxidation of glucose. The potential energy of the glucose is passed from compound to compound and 70% of it appears as ATP. The grey rectangles represent the approximate amounts of energy reaching the various compounds. (Figure and legend from Lehninger, *Sci. Am.* **202**, No. 5, 102, 1960.)

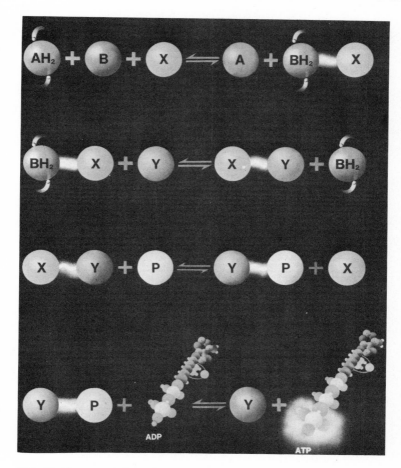

FIG. 45 Representation of presently accepted view of how transfer of electrons is coupled to the phosphorylation process. A reduced "transferrer" AH_2 reacts with the other "transferrer" B_1 in the presence of compound X, reducing B to $BH_2 \sim X$. The symbol $\sim$ indicates a high energy bond. In this scheme electrons are depicted as small white spheres with tails rotating around AH_2 and BH_2. It is the transfer of these electrons which constitutes the reduction process. At this point $BH_2 \sim X$ reacts with another substance, Y, forming a high energy level intermediate $X \sim Y$. At this stage the inorganic phosphate P intervenes, linking to Y—still through a high energy bond $\sim$ and releasing X. $Y \sim P$ reacts with ADP to form ATP + Y. The energy present in the terminal phosphate group of ATP is indicated by a light halo around it. (From *Rassegna Med.* **42**, No. 2, 1965.)

FIG. 46

Beautiful studies using special high-resolution techniques in electron microscopy by Dr. H. Fernandez-Moran and the studies of biochemists, e.g., Dr. B. Chance, have suggested that minute particles are attached to both sides of the mitochondrial membranes and that the energy-producing electron-transfer chain extends from the particle on one side across the membrane to the particle on the other side. (This is demonstrated graphically in Figs. 41 and 42.) The work of Green and Lehninger and their colleagues have shown that the chemical activities taking place in the Krebs cycle (see below) release electrons that via groups of oxidation—reduction proteins known as complexes 1, 11, and 111 and via coenzyme Q, are passed to cytochrome C and via complex IV to combine with oxygen with the production of water. These complexes are located within the lipid layer of the membranes together with coenzyme Q and cytochrome C. It is as a result of the reactions taking place in this electron-transport chain that ATP is produced, each pair of electrons transferred in fact producing three molecules of ATP.

In the spheroidal particles on the one side of and directly attached to the membrane, the Krebs cycle is situated. The operation of the cycle can be initiated by feeding into it molecules of acetyl coenzyme A, which can be derived from pyruvic acid derived from lactic acid or from lipids or amino acids. The electrons derived from the activity of the Krebs cycle pass to the mitochondrial membrane proper where they may be taken up by complex I or by complex II. Those that go to complex I produce ATP at that site and are then passed to coenzyme Q (which, incidentally, is chemically related to vitamin K). Electrons from coenzyme Q are passed into spherical bodies attached to the mitochondrial membrane by a short stalk. In the stalk is complex III where another molecule of ATP is produced. The electrons are then handed on via cytochrome C to complex IV (both cytochrome C and complex IV are in the spherical part

---

FIG. 46 The process of glycolysis is demonstrated by this figure. In stages 1 and 3 two molecules of ATP are used up, but four more are formed in stages 8 and 11. The net gain is two molecules of ATP. 1. Glucose, 2. glucose-6-phosphate, 3. fructose 6-phosphate, 4. fructose 1,6-diphosphate, 5. 3-phosphoglyceraldehyde, 6. dihydroxyacetone phosphate, 7. 1,3-diphosphoglyceric acid, 8. 3-phosphoglyceric acid, 9. 2-phosphoglyceric acid, 10. enolphosphopyruvic acid, 11. pyruvic acid, 12. lactic acid. (From *Rassegna Med.* **42**, No. 2, 1965.)

of the stalked body) from where they are attached to oxygen while another molecule of ATP is born. Electrons may however proceed to complex II and from there to coenzyme Q—no ATP is produced by that path but these electrons handed on by coenzyme Q have now followed the same route to oxygen as those which go first to complex I. The production of ATP by mitochondria results in an energy yield of 50% as compared with that of a modern steam engine—8% (see Figs. 43–46).

It is of interest that in a single 24-hr period the human body forms, and destroys, about its own weight in ATP (70 kg).

According to Green, complex I contains DPNH (now known as

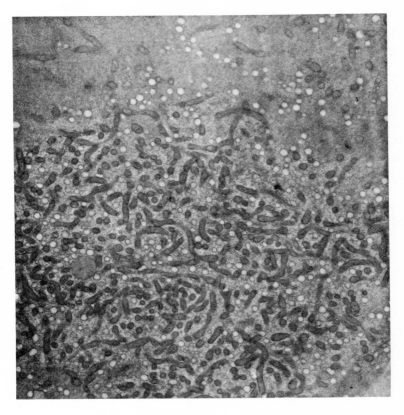

FIG. 47 Mitochondria aggregated in the region of yolk production in egg of newt. (Electron micrograph by J. Hope.)

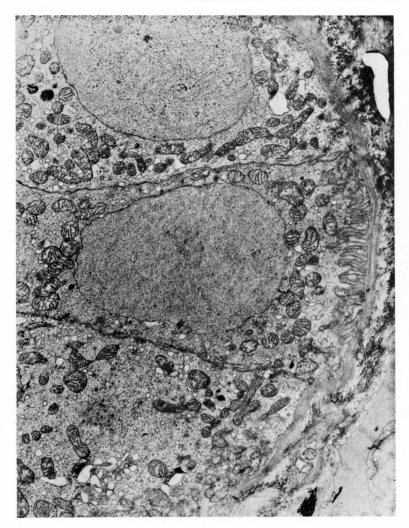

FIG. 48 Mitochondria in tubule cells of rhesus monkey kidney. (Electron micrograph by J. Hope.)

NADH or reduced nicotinamide adenine dinucleotide) + cytochrome reductase; complex II contains succinic co-enzyme Q reductase; complex III, QH_2, cytochrome reductase; and complex IV,

$$\underbrace{\text{reduced cytochrome C, oxygen reductase}}_{\text{cytochrome oxidase}}$$

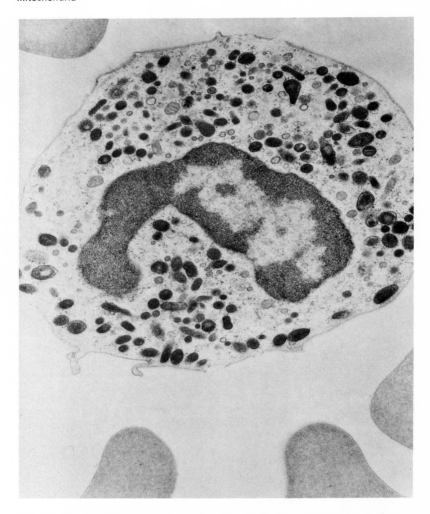

FIG. 49 Neutrophil polymorphonuclear leucocyte from Chimpanzee. Note negligible endoplasmic reticulum, few mitochondria, many dark bodies which are the neutrophilic granules. Magnification 24,000×. This cell does not synthesize protein. (Electron micrograph by C. Webb.)

Although this explanation of the path of the electron-transfer chain in mitochondria seems neat and tidy, it must be admitted that there it has been subjected to some criticism. It must be remembered that we are dealing here with a problem that is right on the edge of advancing knowledge and the techniques for studying it are at the limit of their capability. Scientists such as B. Chance, V. P. Whittaker, and E. Racker have drawn attention to some inconsistencies between this conception of the localization of the electron chain in mitochondria. One of the criticisms was that the particles attached to the mitochondria were too small for the entire respiratory

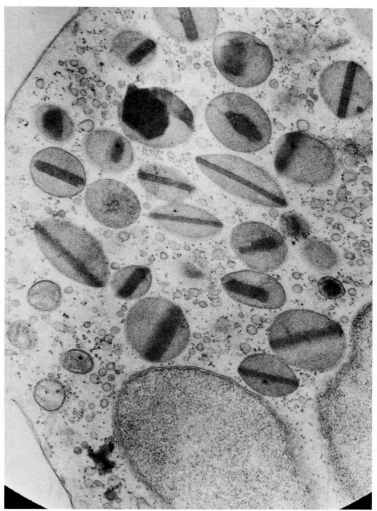

FIG. 50 Eosinophilic leucocyte of Orangutan. Nearly all large eosinophilic granules have "bar" running across center. The nature and function of this bar is unknown. Small number of mitochondria negligible endoplasmic reticulum. Magnification 45,000✕. (Electron micrograph by C. Webb.)

assembly to be contained in them. This criticism has been side-stepped as shown in the diagram where parts of the respiratory assembly are in the particles in one side of the membrane and others in the particles on the other side of this membrane and a significant part of the electron chain actually passes along the

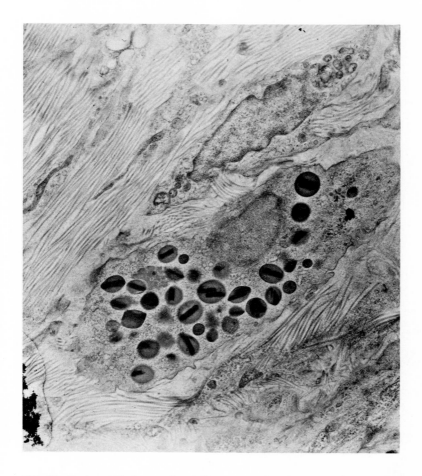

FIG. 51 Eosinophilic leucocyte of Rhesus monkey in connective tissue of ovary. The "bars" in the eosinophilic granules are very well demonstrated. The leucocyte is surrounded by many cross banded fibrils which are composed of collagen.

mitochondrial membrane itself. One important piece of work against the hypothesis outlined above is that by Chance, Parsons, and Williams [*Science* **143,** 136 (1964)], in which they describe how they stripped all the particles from the mitochondrial membranes and yet these membranes retained their cytochromes and Stasney and Crane [*J. Cell Biol.* **22,** 49 (1964)] found that when they studied various submitochondrial fractions separately the fraction richest in particles from the mitochondrial membrane had the least respiratory activity. As shown in the diagram, a considerable part of electron-transport chain is in the mitochondrial membrane, and one would therefore not expect isolated particles to do their respiratory work properly in an isolated condition. In this stratospheric area of fine techniques, it is difficult to eliminate from anybody's work the possibility that technical error might be responsible in part or wholly for the results they obtain. Although the possibility of the stalked particles being artifacts has been suggested, no evidence has been brought forward to show that this is so, and anyone who knows the superb technical excellence of Dr. Fernandez-Moran and his extensive experience in interpreting such microstructural phenomena will find it difficult to accept these structures as artefacts. In the meantime, therefore, the scheme described provides a reasonable working hypothesis concerning the localization of the respiratory assemblies on the different parts of the mitochondria. See Figs. 47–51 for mitochondria in different cells.

PLASTIDS AND CHLOROPLASTS

Plant cells contain a number of plastids some of which are used for storage; e.g., in the potato (the tuber), plastids are used for the storage of starch. Sometimes plastids contain oil and sometimes protein. Some plastids, however, contain chlorophyll and are important metabolic organelles. It is of interest that biochemists thought at one time that it ought to be possible for solutions of chlorophyll to have the same synthetic activity (starch production) as chloroplasts, but since then it has been demonstrated that the chlorophyll molecules have to be arranged in a regular fashion before they can perform this function and that in chloroplasts this regular structure is present. There are usually between 20 and 40 chloroplasts in the cells of higher plants. These chloroplasts are usually biconvex and measure about 5 μ by 3 μ. Lower plants, e.g., algae, may contain

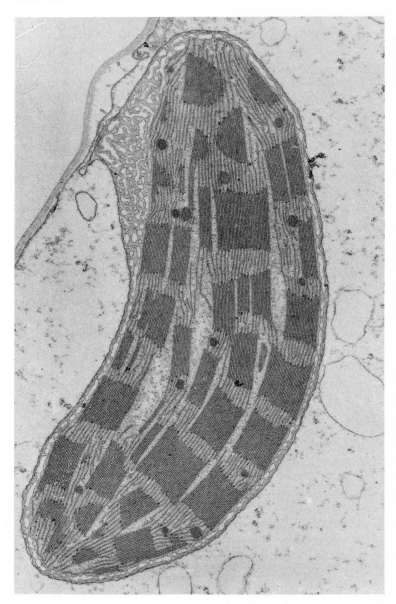

FIG. 52 Section of chloroplast from maize plant showing grana and stromal lamellae (see text) 32,000×. (From E. J. DuPraw, "Cell and Molecular Biology," Academic Press, 1968.)

far fewer and often exotically shaped chloroplasts. In the case of the alga *Spirogyra,* a single helically arranged chloroplast is present that often contains large granules of starchlike material arranged in "pyrenoids." It is of interest that in the cells of higher plants where the chloroplasts are more or less motionless they will still turn toward the source of light or even change their shape to get the maximum advantage from the light. If the chlorophyll-containing protozoan *"Euglena"* is kept in the dark, its chloroplast disintegrates, but if it is replaced in the light the chloroplast will re-form. Thus light can itself induce the formation of a structure designed to utilize it if the raw materials are present within the cell; chloroplasts can also change their volume under the influence of light.

Meyer (1893) is usually given the credit for the discovery of chloroplasts. He even described them as being composed of filaments (grana) surrounded by a material described as "stroma."

Electron microscope studies of chloroplasts made in the early 1950's confirmed that they contained filaments that were in fact made up of a series of lamellae. Some of these lamellae stain with osmium tetroxide and alternate with nonosmophilic bands. These bands are about 65 Å thick. The chloroplast itself is surrounded by a membrane about 100 Å thick, which is made up of two of Robertson's "unit membranes," each about 50 Å thick. The chloroplast contains a "stroma," which is a sort of matrix with an affinity for water and is liquid in nature. Enclosed and surrounded by the stroma are a series of Meyer's "grana," which are lamelliform in shape. The "grana" may number as many as 80 and measure about 0.5 μ in diameter (see Fig. 52). Some lamellae extend into the stroma between the lamelliform groups of grana and are known as "stromal lamellae." The walls of both the lamellae and the grana are composed of both protein and phospholipid.

Park and his colleagues, in a series of interesting scientific papers, have shown a series of particles on the surface of each disk (lamella), which have been called quantasomes. Each quantasome has been reported to measure 187 × 155 × 100 Å and to be composed of four smaller units. The quantasomes contain both protein and lipid (see Fig. 53). Among the compounds found in the lipid part of quantasomes are chlorophylls, carotenoids, quinone compounds, phospholipids, and sulfalipids. In the protein fraction cytochrome and various enzymes are present. The chlorophyll molecules appear to be attached to phospholipid molecules, which are them-

FIG. 53 A paracrystalline array of quantasomes. Taken from chloroplasts of spinach. (From R. Park, Chloroplast Structure in Photosynthesis, *Intern. Rev. Cytol.* **20,** 1966.)

selves separated by molecules of carotenoid pigment. Underlying this complex is the protein layer that contains the enzymes producing oxygen and on the other side, in the stroma, are protein-containing enzymes reducing CO_2. The ATP provides the energy for subsequent synthetic reactions involved in the formation of starch.

Green and Goldberger have pointed out ("Molecular Insights into the Living Process," Academic Press, 1967) that there is a fundamental similarity between mitochondria and chloroplasts.

The basic ultrastructural patterns are virtually the same—the mitochondrial "elementary particle" is duplicated by the "quanta-some" of the chloroplast (see Fig. 54). The electron-transfer chain appears to be more or less identical in the two organelles and it is coupled to ATP synthesis. There are a number of other points of similarity.

It is of interest that isolated chloroplasts can carry out the photosynthetic cycle, involving the liberation of oxygen and the fixation of CO_2 as sugar and starch and the production of ATP from ADP by phosphorylation.

The reactions concerned with electron transport in the chloroplast that proceed from water to the reduction of ferredoxin take place in the membrane system in the light, whereas the actual fixation of CO_2 occurs in the stroma and takes place in the dark with the help of a key enzyme known as carboxydismutase. In the latter process, carbon dioxide condenses with ribulose diphosphate and together with hydration this procedure produces two molecules of phosphoglyceric acid. This is the particular reaction which is catalyzed by carboxydismutase and the reactions form a part of a cycle known as the "Calvin cycle." Each complete reaction involves six turns of the Calvin cycle, this results in six molecules of carbon dioxide being bound to six molecules of ribulose diphosphate and the production of 12 molecules of phosphoglyceraldehyde. This procedure uses up the energy in 12 molecules of ATP. Ten of the phospho-glyceraldehyde molecules are used to regenerate six molecules of ribulose diphosphate (which uses another six molecules of ATP) so that the cycle can be set going again, the remaining molecules of phosphoglyceraldehyde combine to form a molecule of glucose (see Fig. 55).

The reaction can be summarized as follows:

$$6\,CO_2 + 12\,NADPH + 12\,H^+ + 18\,ATP + 12\,H_2O \rightarrow C_6H_{12}O_6 + 12\,NADP + 18\,ADPH + 18\,POH.$$

It will be noted that the oxidation of reduced nicotinamide adenine dinucleotide with three phosphate groups (NADPH) is an essential part of this reaction. This compound was formally called triphosphohyridine nucleotide or TPN. The production of reduced

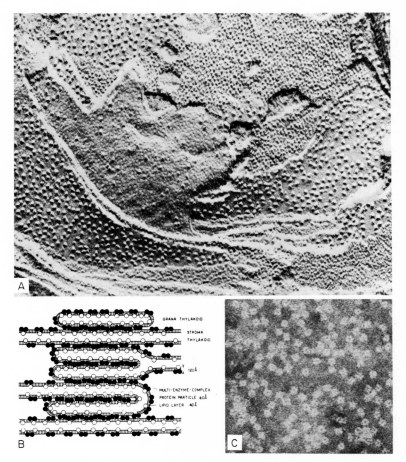

FIG. 54 (A) Use of freeze etching to reveal chloroplast structure. Different types of lammelar surfaces are shown. On surfaces are many small closely packed particles. Magnification 90,000×. (From D. Branton and R. Park, *J. Ultrastructure Res.* **19,** 283, 1967.) (B) Diagrammatic representation of distribution of multienzyme complexes, protein particles, lipid layers, etc., as revealed by freeze etching by K. Muhlethaler. [*In* "Biochemistry of Chloroplasts" (T. W. Goodwin, ed.), Vol. I, p. 49, Academic Press, 1966.] (C) Granules of the enzymes which catalyze condensation of CO_2 and ribulose diphosphate–carboxydismutase. Particles are 80 × 111 Å. Magnification 200,000×. [From R. Park, *in* "The Chlorophylls" (L. P. Vernon and G. R. Seely, eds.), Academic Press, 1966.] (This plate combining the three figures is taken from E. J. DuPraw, "Cell and Molecular Biology," Academic Press, 1968.)

pyridine nucleotide is part of the photosynthetic reaction which takes place in the light. This is known as a "noncyclic mechanism." In this process electrons are drawn from a donor believed to be the OH ions normally present in such an aqueous medium, this releases oxygen, long recognized as a product of green plants exposed to light. An intermediate substance in this electron flow is the iron-containing compound "ferredoxin." It is first reduced by the electrons derived from the OH ions and then transfers its electrons to NADP resulting in its reduction to NADPH. The enzyme responsible for this is (ferredoxin–NADP reductase). ATP is produced during the light period of photosynthesis by a process known as "cyclic photophosphorylation." The reduction of ferredoxin occurs in this process. Electrons flow from chlorophyll to ferredoxin and then return to chlorophyll via a transparent chain that includes cytochromes and plastoquinone (related to vitamin K). During this process coupled phosphorylation reactions of an unknown nature produce ATP.

The NADPH and the ATP formed in this way in the membranes under the influence of light thus become available for the chemical reactions involved in fixing CO_2 into glucose and starch that takes place in the dark in the stroma.

Du Praw ("Cell and Molecular Biology," Academic Press, 1968) compared the quantasome, where chlorophyll and other pigments are located, with "a tiny photoelectric cell coupled to a storage battery." The light energy produces a flow of electrons and these are stored as energy in the storage battery as ATP.

LYSOSOMES

In the first studies on differential centrifugation of cell particles, the acid phosphatase activity of the cell was found to be contained in the mitochondrial fraction; however, if the force of centrifugation was not quite so strong the acid phosphatase activity was found to be present in the microsome fraction thus indicating that this enzyme was contained in a body of intermediate density. Subsequently it was found by De Duve that these particles also contain cathepsin (a proteolytic enzyme), acid ribonuclease, acid desoxyribonuclease, and B glucuronidase. De Duve described these particles as "lysosomes" and referred to them as "suicide bags," since in a damaged or dead cell the enzymes would be discharged from the bodies that contain them and would autolyze the cell.

FIG. 55 Idealized diagram of distribution of multienzyme complexes, etc., in chloroplast lamellae to show distribution of function in structure and the mechanism of ATP production. Small streaks of gray ending in gray spheres (F) represent photons (F) coming from the sun. They remove an electron from the chlorophyll (C), which is captured by an acceptor unit (Vitamin K or flavine-adenine dinucleotide (1). The electron is then transferred to a second acceptor chain (2) (cytochromes). During the transfer from (1) to (2) the electron loses energy with the production of an ATP molecule. From (2) the electron is passed back to the original chlorophyll molecule from which it came. In this passage another molecule of ATP is formed. ATP is also produced in another way in the chloroplast, shown to the right of the system lust described. Water absorbed by the plant is ionized H^+, OH, and electrons.

Other enzymes have since been discovered in lysosomes. These are: phosphoprotein phosphatase, B-galactosidase, glucosidase, mannosidase, arylsulfatase A and B, B-N-acetylglucosamidase. It appears however that there is a range of lysosomes and not all the 12 enzymes described above are found in any one lysosome—it seems unlikely, however, that there is one type of lysosome for each enzyme.

Histochemical preparations for acid phosphatase in intact cells have often demonstrated that the enzyme appears to be contained in discrete particles, and sedimentation studies have confirmed that these particles are about 0.4 μ in diameter and have an average density of about 1.15. If the particles are treated with lecithinase or proteolytic enzymes (which break down membrane structure), the enzymes escape; this suggests that the lysosomes are surrounded by a lipoprotein membrane.

In 1956, Novikoff and De Duve made electron-microscope studies of pellets of sedimented material which had been shown to be rich in lysosomal enzymes and found a number of particles to be present. These particles were found to be bounded by single membranes (a "unit membrane," Robertson, 1959). There appeared to be a central structureless cavity in these bodies. Some of these bodies appeared to contain granules of ferritin. A study of electron micrographs of sections of intact tissue showed bodies, similar to those found in the sediment, present in the cytoplasm of cells. Eventually Novikov and his colleagues showed by electron histochemistry that these electron-dense bodies contained acid phosphatase. Studies of many organs and tissues showed that lysosomes were present in the cells of all those examined. The from the Golgi apparatus, and others believe that they have their origin of lysosomes is uncertain. Some workers believe they originate origin in the smooth endoplasmic reticulum. There is some evidence that at least some pinocytotic vesicles fuse with lysosomes.

$2H^+$ reduce the hydrogen acceptor TPN to PNH^+. Two electrons released from a chlorophyll molecule by photons are captured by the $TPNH^+$ which is tranformed into $TPNH_2$. The electrons supplied by the water are captured by an acceptor chain (cytochromes) (3) and returned to the two chlorophyll molecules which gave up electrons to TPN. During this latter process the electrons lose energy with the production of two molecules of ATP. The ATP is used to supply energy for the various synthetic processes which take place in the chloroplast. (From *Rassegna Med.* **42,** No. 2, 1965.)

De Duve has pointed out that it is of special interest that such hydrolytic enzymes are in special particulate bodies differing from others of the cytoplasm. He is not clear as to how this should be interpreted but has pointed out that, if hydrolytic enzymes are free to act within the living cell, as, for instance, they can do in homogenates, they would interfere with the efficiency of the synthesis and may even affect the structural integrity of the cell; he then suggests that segregation of hydrolytic enzymes in this way is one method by means of which this activity is either kept in check or localized in specific parts of the cell.

A number of studies have made it very likely that lysosomes are related to cell damage and recent work in England strongly suggests that arthritis is due to damage inflicted on cartilage by enzymes leaking from the lysosomes of the cartilage cells. The present author found an increase of hydrolytic enzymes in senescent tissues and suggested that in old tissues these enzymes may be less effectively contained in the lysosomes and that this may be partly responsible for cellular aging. Some authors believe that the hormone cortisone from the adrenal gland stabilizes the lysosome membrane and that this would explain the efficacy of cortisone in the treatment of rheumatoid arthritis (see Fig. 56).

Another set of particles has more recently been identified in cells, they are known as "microbodies" or "peroxisomes." These particles have a single membrane, but it is thinner than that of the lysosome. The interior of the particle contains a granular matrix and in some species of animal there is a core that is more electron dense than the matrix. These bodies are found in the cell in association with the smooth endoplasmic reticulum and they contain a number of enzymes, which include catalase and one or more hydrogen-peroxide-producing oxidases. De Duve has found recently that peroxisomes contain the key enzymes of the glyoxylate cycle, one of the pathways for carbohydrate metabolism. De Duve believes that one of the functions of peroxisomes is to play a part in gluconeogenesis—the formation of glycogen or glucose; he believes that they also play a part in the reoxidation of nicotinamide adenine dinucleotide (NAD) a respiratory coenzyme. There is a possibility that the peroxisome has evolutionary importance since they may represent primitive respiratory particles whose functions have been largely replaced by mitochondria.

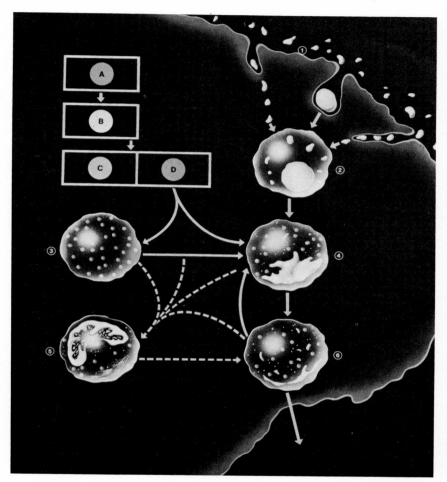

FIG. 56 The lysosomes are edged by a membrane and contain enzymes that demolish large molecules (lipids, proteins, etc.) into fragments, which may then be oxidized by the mitochondria. Lysosomal enzymes (acid hydrolase) (B) are synthesized in the ribosomes (A) and then stored in Golgi's apparatus (C) in the form of tiny vesicles (D) that are changed into "primary lysosomes." The foreign particles, which the cell engulfs by endocytosis, are surrounded by a membrane and form phagosomes, which in turn merge with lysosomes and give rise to secondary lysosomes of digestive vacuoles. The unused material is left in the residual bodies that may either remain in the cell or eventually expel their contents outside the cell membrane. 1. Cell membrane, 2. phagosome, 3. primary lysosome, 4. digestive vacuole, 5. autophagic vacuole, 6. residual body. (From *Rassegna Med.* **45,** No. 3, 1968.)

THE METABOLISM OF CARBOHYDRATES

The role of mitochondria in oxidative phosphorylation has already been mentioned and their role in the metabolism of carbohydrates through the presence in their substance of enzymes concerned with the Krebs cycle and the cytochrome system has been indicated. We should now consider the problem of carbohydrate metabolism and the role that mitochondria and other parts of the cytoplasm play in it.

Carbohydrate metabolism is extremely important for cell synthesis and is the main source of energy for cell activities. Before attempting to localize the various activities of carbohydrate metabolism in the actual parts of the cell, we should consider briefly what the metabolism of carbohydrates involves. There are two types of metabolism, anaerobic and aerobic. The anaerobic route is demonstrated very well by muscle, and most of the information on this type of metabolism of carbohydrates has been obtained by studies of this tissue.

The result of anaerobic metabolism is the production of lactic acid and the liberation of a good deal of CO_2. However, although we think in terms of anaerobic metabolism for muscle, we have to realize that muscle itself has a first-class blood supply that appears to be increased by various physiological mechanisms when muscle is forced to do work and that muscle in the process of contraction uses a rather surprisingly large amount of oxygen.

Lactic acid has been shown to accumulate in muscle extracts and in isolated muscles kept under anaerobic conditions. If we consider the accumulation of lactic acid in an animal *in vivo*, we find that after moderate work the accumulation of lactic acid goes up slightly but remains at a pretty steady level. However, if strenuous work is done, then the amount of lactic acid goes up extremely steeply and slowly comes back to normal after the work has ceased. The reason for this is that under normal circumstances muscle can obtain oxygen fast enough to reoxidize the lactic acid as rapidly as it is formed and only a small amount of lactic acid accumulates. However, it is possible for muscle to do more work than it can supply oxygen for and it can do this by oxidizing carbohydrates anaerobically and so accumulating lactic acid. Eventually this lactic acid has to be oxidized with the aid of atmospheric oxygen but it can be stored and reoxidized later and this can take place over a longer period of time. Some of the lactic acid is converted into glycogen in the liver and the rest is oxidized. The fact that it is possible to accumulate lactic acid and slowly oxidize this after the work is finished provides a mechanism by means of which an "oxygen debt" can be produced. Under aerobic circumstances it is not lactic acid that is formed in the metabolism of carbohydrates but pyruvate. However, this does not accumulate and it is oxidized almost as rapidly as it is formed— as we shall see in a minute there is a very complicated system for oxidizing this pyruvate. The only time when pyruvic acid does accumlate in the tissues is in the absence of vitamin B_1 or thiamine, a fact that was demonstrated years ago in Oxford University by R. A. Peters. Under a process known as oxidative decarboxylation with the aid of the cocarboxylase (thiamine pyrophosphate), it yields acetate, carbon dioxide, and lactate. It has been stressed that this reaction is fundamentally of an oxidative nature and is called *oxidative decarboxylation,* an important process both in carbohydrate and protein metabolism.

We have mentioned the production of pyruvate but have not yet considered the process by means of which this compound is produced—this process is known as glycolysis and it represents the preliminary stage in the metabolism of carbohydrates. Glycolysis is, in effect, the reverse of photosynthesis. In photosynthesis the energy of sunlight is used to combine CO_2 and water into carbohydrates; in the process of glycolysis, the glucose that is formed from carbohydrates such as glycogen and other polysaccharides is converted into pyruvate and then into CO_2 and water with the liberation of energy. We can express this as $C_6H_{12}O_6 + 6O_2 = H_2O + 6CO_2 +$ energy. This oxidation of glucose to give CO_2, water, and energy is not a single step but involves a very large number of steps of considerable complexity. During the various steps, small packets of energy are released at a rate at which the cell can use them, whereas if there were a sudden explosive release of energy by the oxidation of glucose, the cell would probably not be able to use this relatively large amount of energy in a coordinated way and a good deal of it would probably be wasted.

The first stages of glycolysis involve the phosphorylation of glucose and this is done with the aid of ATP as follows (the enzyme concerned in this process is placed above the arrow):

$$\text{glucose} + \text{ATP} \xrightarrow{\text{hexokinase}} \text{glucose-6-phosphate} + \text{ADP}$$

Hexokinase is not just one enzyme, there are in fact a number of hexokinases that catalyze phosphorylation of hexoses. The phosphorylation of glucose results in the transferring of a high-energy phosphate from the ATP to glucose and so to form a phosphate ester that is poor in energy—this type of reaction is called an exergonic reaction and is essentially irreversible. It is of interest that one of the properties of glucose-6-phosphate that differs from glucose is the fact that the phosphate ester has difficulty in penetrating cell membranes whereas glucose itself, apparently, crosses without any difficulty, and it has been suggested that this hexokinase reaction is one way in which glucose can be locked in a cell. It is also an essential prerequisite for the resynthesis of glycogen.

Many things can happen to glucose apart from phosphorylation and ultimate conversion into CO_2 and water via the glycolytic and Krebs cycle system. Among these are its dehydrogenation by a glucose dehydrogenase to form gluconic acid. This is done by a diphosphopyridine nucleotide (DPN)-linked (codehydrogenase) reaction.

An International Commission has recently altered the names of some of these coenzymes—for example, DPN now becomes nicotinanide adenine dinucleotide (NAD); when it has an additional third phosphate group, it is known as NADP, which was called triphosphopyridine nucleotide (TPN) before the Commission changed its name. This compound, though closely related to NAD, cannot replace it in the reactions in which it forms an integral part. Likewise, NAD cannot replace NADP in the reactions in which it is a constituent. NADP seems to play an electron-transport role in a smaller number of dehydrogenases than NAD. Glucose-6-phosphate dehydrogenase is a typical enzyme for which NADP acts as an electron acceptor. The reduced forms of these coenzymes are indicated by NADH and NADPH.

In mammals, it is believed that this is not a usual pathway for glucose to follow. If glucose can be locked into position in a cell by being converted into a phosphate, it is obvious that there must be in existence a mechanism that can release it again since it needs, for instance, to be fed from the liver periodically into the bloodstream to keep the blood glucose level at a relatively constant figure. This is carried out by a specific enzyme, glucose-6-phosphatase,

$$\text{glucose-6-phosphate} + H_2O \xrightarrow{\text{G-6-Pase}} \text{glucose} + PO_4.$$

This glucose-6-phosphatase is probably present in all tissues which release glucose from cells, but it does not appear to occur in skeletal muscle.

$$
\begin{array}{ll}
\text{HCOH} & \\
| & \\
\text{HCOH} & \\
| & \\
\text{HOCH} & \quad\text{O} \\
| & \\
\text{HCOH} & \\
| & \\
\text{HC} ———— \\
| & \\
\text{H}_2\text{COPO}_3\text{H}_2 &
\end{array}
\qquad
\begin{array}{ll}
\text{HCOPO}_3\text{H}_2 & \\
| & \\
\text{HCOH} & \\
| & \\
\text{HOCH} & \quad\text{O} \\
| & \\
\text{HCOH} & \\
| & \\
\text{HC} ———— \\
| & \\
\text{H}_2\text{COH} &
\end{array}
$$

Glucose-6-phosphate $\qquad$ α-Glucose-1-phosphate

Glucose-6-phosphate may be converted into glucose-1-phosphate, in other words the phosphate is simply shifted around from the 6-position to the 1-position on the glucose molecule. This change of position of the phosphate group is catalyzed by an enzyme known as phosphoglucomutase, and the reaction changing the phosphate group from one part of the molecule to the other is an easily reversible reaction. The formation of glucose-1-phosphate from glucose-6-phosphate is a stage in the synthesis of glycogen, for instance, if glucose is accumulating in a cell it is converted to glucose-6-phosphate and then to glucose-1-phosphate. The molecule of glucose-1-phosphate then polymerizes into glycogen. The reverse process can occur—glucose-6-phosphate can be obtained from glycogen by first producing glucose-1-phosphate and then converting it into glucose-6-phosphate. We have, however, diverted a little from the direct line of our story of the glycolytic cycle. After the formation of glucose-6-phosphate from glucose, the next stage in the glycolytic cycle is the conversion of glucose-6-phosphate into fructose-6-phosphate. This reaction is readily reversible and is catalized by an enzyme called phosphohexose isomerase.

Fructose-6-phosphate may also be formed directly from fructose, and it is known that an enzyme, fructokinase, which is present in brain and muscle can produce this effect. The next step is the further phosphorylation of fructose-6-phosphate; another phosphate group is added in the 1-position to give fructose 1-6-diphosphate. The enzyme responsible for this is phosphofructokinase and the reaction is carried out with the aid of ATP:

$$\text{fructose-6-phosphate} + \text{ATP} \xrightarrow{\text{PFkinase}} \text{fructose-1-6-diP} + \text{ADP}$$

This reaction is exergonic (heat producing).

Fructose-1-6-diphosphate is also known as hexose diphosphate and in the next stage this is broken down by what is described as the "aldolase" reaction into three phosphorylated compounds— triosephosphates. The first of these is ketose triosephosphate, the second is dihydroxyacetone phosphate, the third is phosphoglyceraldehyde. The triosephosphate can be converted into phosphoglyceric acid; dihydroxyacetone phosphate can be converted into phosphoglyceraldehyde or alternatively it can be reduced to α-glycerophosphate with the aid of DPNH (NADH) and α-glycerophosphate dehydrogenase. This reaction is potentially important for the synthesis of lipids since from the α-glycerophosphate, phosphatidic acid can be synthesized, and phosphatidic acid can be the starting point for the synthesis of lecithin, cephalin, and also of fats. Phosphoglyceric acid, with the aid of the enzyme enolase, becomes converted into phosphoenol-pyruvic acid.

Pyruvic acid penetrates cell membranes very well and can thus leave the cell and in theory can be distributed to any cell in the body.

Conversely, all the steps we have mentioned can be reversed and pyruvate can be converted back into glucose-6-phosphate. When pyruvic acid is reduced it gives lactic acid. Lactic acid itself can be converted back to pyruvic acid. Liver cells are capable of reversing the whole glycolytic series of reactions and can produce glucose and glycogen again from lactic acid. Muscle can reverse lactic acid to glucose-6-phosphate and glucose-1-phosphate and glycogen but cannot produce nonphosphorylated glucose.

In these first stages of carbohydrate metabolism, a certain amount of energy is liberated but this is not much more than about one-tenth of the total amount of energy which is produced by the production of CO_2 and water from glucose. The glycolytic part of the cycle is the least energy-producing part.

In addition to its conversion into lactic acid or its oxidation, pyruvic acid can be converted to alanine, an amino acid, by the process of transamination. Thus here one can see a link between carbohydrate and protein metabolism.

If there is ample oxygen, the pyruvic acid produced by this

first stage of carbohydrate metabolism can be oxidized. This is a complex process in which in the first stage the pyruvic acid is converted by the process of oxidative decarboxylation into acetyl coenzyme A and CO_2. Thiamine pyrophosphate (cocarboxylase) is an essential enzyme for this process.

Acetyl coenzyme A is a two-carbon substance and it condenses with oxaloacetic acid which is a four-carbon dicarboxylic acid to yield a six-carbon tricarboxylic acid, namely, citric acid (or citrate). The enzyme catalyzing this is called the "condensing enzyme." Thus starts a series of changes that ultimately lead to the formation of CO_2 and water. The production of citric acid is followed by the loss and recapture of water, and it becomes converted to cis-aconitic acid or cis-aconitate (with the aid of aconitase, glutathione, and ferrous iron), which on further hydration is converted into isocitric acid. This production of isocitric acid can, however, take place directly without the production of free cis-aconitate (cis-aconitic acid). This compound then loses hydrogen and thereby becomes oxidized to oxalosuccinic acid or oxalosuccinate (this reaction is catalyzed by isocitric dehydrogenase) and decarboxylation turns it into α-ketoglutaric acid (the enzyme responsible is oxalosuccinic decarboxylase plus oxidized manganese). α-Ketoglutaric acid is decarboxylated and then oxidized by the loss of two hydrogen atoms to succinic acid (the enzyme concerned is α-ketoglutaric dehydrogenase). More recent studies have shown that this process is more complicated— the decarboxylation and oxidation of α-ketoglutarate is carried out by four coenzymes: thiamine pyrophosphate, lipoic acid, coenzyme A, and nicotinamide adenine dinucleotide (NAD). An intermediary substance succinyl coenzyme A is first produced from α-ketoglutarate and succinate is then liberated from the succinyl coenzyme A. Succinic acid is also oxidized by the loss of H_2 (with the help of succinic dehydrogenase) to fumaric acid. The actual oxidation of succinate occurs by the transfer of two electrons to ferricytochrome B, which is a member of the electron-transport chain. Succinic dehydrogenase is the enzyme that catalyzes this transfer. This enzyme contains iron and a flavin coenzyme. The latter by the addition of the elements of water becomes converted by fumarase into malic acid, and the malic acid by dehydrogenation catalyzed by the enzyme malic dehydrogenase is converted in oxaloacetic acid, and there we are back at the beginning of the cycle. The oxaloacetic acid (oxaloacetate) is ready to combine with another molecule of acetyl coenzyme

A to produce citric acid once more. During this whole process three molecules of CO_2 and five molecules of H_2 are given off.

It can be seen that quite a complex series of reactions take place in what has been called the "tricarboxylic" or "citric-acid cycle," or the "Krebs cycle." All three terms are applicable.

The final combination of the hydrogen atoms liberated by the Krebs cycle with oxygen is brought about by the cytochrome system. It is of interest that cytochrome is a protein that contains a form of heme, an iron containing pigment that is also present in hemoglobin. It is probably more widely distributed than any other type of heme protein since it occurs in the cells of all organisms that use oxygen, irrespective of whether they are animals or plants or whether in the case of animals they are vertebrates or invertebrates, or protozoa.

In the cytochrome system a variety of compounds and enzymes are involved. Cytochrome oxidase is a widely distributed enzyme since its occurrence is comparable in distribution to cytochrome. It has not yet been obtained pure, and it has been found to be bound to the insoluble material when cells are homogenized and spun down. This is because it is associated with the mitochondria. There is also another compound involved called flavoadenine dinucleotide (FAD for short) which is associated with various proteins to form a variety of different enzymes. A flavin containing enzyme was first isolated and called a "yellow ferment" as long ago as 1932 by Warburg and Christian. Flavoproteins are usually associated with various metals; molybdenum, copper, and iron are three that have been found to be necessary for their function. It is not quite clear where the metal atoms are situated on the enzyme molecule. There are two types of riboflavin-containing enzymes, one type is an electron acceptor from reduced DPN (NAD) or TPN (NADP), and one type can transfer electrons either to oxygen or to the cytochromes, whereas the other type of flavoprotein accepts electrons directly from metabolites. Among the important flavoproteins is cytochrome reductase which comes in two types—one for reduced DPN (DPNH) (NAD, NADH) and the other for reduced TPN (TPNH) (NADP, NADPH). The oxidation of these two compounds DPNH and TPNH (NADH and NADPH) in the cell can thus be carried out by the cytochrome system.

Now we are in a position to describe the next stages that

take place in the oxidation of glucose. The five pairs of hydrogen atoms that are passed down to the cytochrome system from the Krebs cycle are combined with the cytochrome and result in its reduction. During this process the hydrogen atoms dissociate to protons and electrons, and the electrons are handed along an electron-transport chain and eventually reduce atmospheric oxygen. The enzyme concerned in the reduction of cytochrome is cytochrome reductase, the flavoprotein already mentioned. Cytochrome is then oxidized with the aid of cytochrome oxidase, which removes the hydrogen atoms. They become combined with atmospheric oxygen to produce water and thus the long journey of oxidation of carbohydrates is done, five molecules of water and three molecules of CO_2 being produced from each molecule of pyruvate. There are, in fact, a series of cytochromes arranged in a chain, and the electrons are passed along these prior to being combined with atmospheric oxygen. Although the course of oxidation may seem tedious and involved to the reader of the preceding pages, all these reactions take place in a flash. Reference back to the section on mitochondria will remind the reader that the electron-transfer chain also goes through coenzyme Q, or *"ubiquinone"* a compound closely related to vitamin K.

An analysis of oxygen uptakes of various tissues in the body gives an indication of the degree to which their cells are metabolizing. As a matter of interest, it might be noted that of the tissues examined, retina and kidney had the highest metabolism and liver was next, then the rate decreased progressively from adrenal, lung, bone marrow, diaphragm, heart, lymph nodes, skeletal muscle, skin to eye lens, which had the lowest level of oxygen consumptions. One important point to remember is that the complex of reactions described above is localized to a great extent in the mitochondria.

One of the important functions of this oxidative cycle just described is its relationship to phosphorylation. We have described earlier the work of Green who showed that oxidative phosphorylation could take place without the whole Krebs cycle occurring, but in intact mitochondria the whole cycle normally goes through and phosphorylation is an important by-product of these reactions. This process of phosphorylation results in the production of high-energy phosphate esters, such as ATP.

NH₂

Adenosine triphosphate (ATP)

Since ATP is one of the principal energy-containing compounds of the body, it is very important in the whole energy cycle of the cell. It is, for example, the main source of energy in musclar contraction and for many of the synthetic processes of other cells.

Although the relation between ATP formation and the respiratory chain has been widely accepted scientifically, Dr. H. Rottenberg of Brooklyn College, at the meeting of the American Society for Experimental Biology held in Atlantic City (1969), has suggested that this relationship is more complicated than it may seem. He described his views as a "chemiosmotic theory" and a number of workers have held similar views. In this theory, instead of cell respiration being directly coupled to phosphorylation (ATP production), the respiratory

FIG. 57 Aldolase preparation of intestine showing diffuse reaction in the smooth muscle. (Figure and legend from Allen and Bourne, *J. Exptl. Biol.* **20,** 61, 1943.)

process generates a proton flux across the inner membranes of the mitochondria and this "proton pump" is geared directly into ATP synthesis.

This briefly is the story of carbohydrate metabolism in the cell. What we want to try to do now is to demonstrate where this complex of activity is situated in the living cell. It appears that most of the processes of respiration and glycolysis actually take place in the cytoplasm and mitochondria. Some years ago (1941), the present author and R. J. Allen demonstrated that zymohexase, which is really a complex of two enzymes and is concerned with the splitting of hexose diphosphate (the aldolase reaction), was localized in the cytoplasm of the cell and in the case of muscle fibers in between the fibrils in the sarcoplasmic material rather than in any of the formed elements. It is noted too that the results of differential centrifugation of cell homogenates have demonstrated that most of the enzymes responsible for the glycolytic cycle have been found to be present either in the supernatant or in "microsomes" and those concerned with the cytochrome system and Krebs cycle are localized specifically in the mitochondria. Now, since the microsomes mostly represent fragments of the endoplasmic reticulum, we can assume that most of the enzymes found in them can also be found in the membranes of the endoplasmic reticulum. In addition to the aldolase complex (see Fig. 57), which has been demonstrated to be present in the supernatant, it has been found that phosphorylase and phosphoglucomutase and glucose-6-phosphate dehydrogenase have also been found in the supernatant and glycolysis has in fact been found to take place in this fluid—a fact that confirms that all the glycolytic enzymes necessary for this process must be present there. However, glucose-6-phosphatase has been found not to be localized at all in the supernatant but exclusively in the microsomes (fragmented endoplasmic reticulum). There is some evidence too that hexokinase is present in the microsomes. On the other hand, DPNH (NADH) and DPN (NAD) and TPN-(NADP-) linked cytochrome c reductases that have been found in high concentration in mitochondria have also been found to be present in the microsomes. It might be possible to suggest from these facts a tentative scheme that would help us to understand to some extent the relationship between the reactions occurring in the cytoplasm itself and those in the mitochondria.

Let us start off by assuming that some glucose has passed

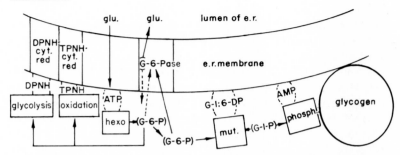

FIG. 58 Schematic representation of relationship of endoplasmic reticulum to carbohydrate metabolism. (From Siekewitz, "Regulation of Cell Metabolism," *Ciba Foundation Symp.,* Churchill, London, 1959.)

across the cell membrane. Perhaps it was taken in by pinocytosis as has been suggested by a number of authors or, possibly, if and when the membranes of the endoplasmic reticulum are continuous with the outside of the cell, it may simply have passed up the cisternae between the membranes of the reticulum and entered into the cell through the membranes of this reticulum. Let us assume that the glucose has passed into the cell by the process of pinocytosis. Then, with hexokinase present in the cytoplasm, it can be converted into glucose-6-phosphate and subsequently run through the cycle to pyruvate. At this point the enzyme system associated with the mitochondria comes into play. If the glucose has to pass across the cell membrane or across the membranes of the endoplasmic reticulum, it should also be able to do this without too much difficulty. If any glucose-6-phosphate should occur in the cisternae, it will probably be able to pass freely into the cytoplasm because of the presence of glucose-6-phosphatase in the membrane. It may be that there is hexokinase in the fluid within the cisternae and that, perhaps, all the glucose there (if any) is first phosphorylated and then released in a timed fashion into the cytoplasm through the action of glucose-6-phosphatase in the membranes (see Fig. 58). Perhaps this is a way of controlling the feeding of the carbohydrate fuel into the furnace. It will be remembered that glucose-6-phosphate passes cell membranes with difficulty unless there is the appropriate hydrolytic enzyme on the membrane. The glucose, once into the cytoplasm by whatever route, can be synthesized into glygogen or it can be brought to pyruvate by the enzymes of the glycolytic cycle that are known to occur in the cytoplasm. Once pyruvate is prepared, it is changed into acetyl coenzyme A and is ready for the next stage. For this it has to come into contact with the mitochondrial membrane. Now, it has been mentioned that a great proportion of the enzymes concerned with the Krebs tricarboxylic acid cycle are localized in the mitochondrial membranes or particles associated with them and since acetyl coenzyme A appears to be the most commonly used fuel of this system, it is at this point that the mitochondria largely take over the final oxidative stages. Thus, as far as we can tell at the moment either in the cytoplasm of the cell itself or in the endoplasmic reticulum, glycolysis occurs with the production of either acetyl coenzyme A or pyruvic acid and then these compounds are converted by the mitochondrial enzymes into CO_2 and water, thus completing the oxidation of carbohydrates. We do not know at what stage these compounds are fed to the mitochondria. They must go largely to the mitochondria because these organelles contain the greatest proportion of the Krebs cycle enzymes—it is these latter that complete the oxidative breakdown of carbohydrate that was set into motion by hexokinase. If under normal conditions we have a large number of mitochondria occupying the cell, we can assume, other things being equal, that oxidation should be proceeding at a fast rate because of the large mitochondrial surface area available. However, where one gets a large increase of number of mitochondria as, for example, in starvation, this may be compensatory hypertrophy and not indicative of increase in oxidative activity. We do not know whether the nature of the mitochondrial surface varies from time to time but we know that the surface area varies. Mitochondria, for instance, fragment from the filamentous and rodlike condition to the granular state under a variety of circumstances. These include mechanical damage to the cells, rough handling, influence of bacteria and other toxins, as a result of anesthesia and anoxia, and so on. It can be shown mathematically that there is a greater surface area available the more fragmented the mitochondria become. Because of this either the pyruvate and/or acetyl coenzyme A can feed more rapidly into or become attached in larger amounts to the mitochondria when they are in a fragmented condition simply because there is a greater surface area. Thus the mitochondria by virtue of their ability to fragment and re-form into the filamentous condition can act as a throttle controlling the rate of aerobic metabolism in the cell. They may also have an additional means of doing this, a sort of fine control. Since it is fairly certain that enzyme molecules are aligned along the cristae, these structures represent another surface where reactions can take place, and reduction in the size and number of cristae would also have a throttling effect on the metabolism or synthesis due to mitochondria. This can be seen in operation in the case of a mitochondrion that has accumulated a good deal of fat or other product of chemical reactions. In such a case, the cristae are reduced or absent altogether as if metabolic processes are brought to a virtual standstill because of the accumulation of reaction products. This brings us to another process of control, a chemical method known as "feedback," which will be discussed shortly.

Siekewitz believes that both glycogen synthesis and breakdown take place at the surface between the cytoplasm and the ergastoplasmic membranes in association with the enzymes present at those surfaces. He points out, however, that only glucose-6-phosphatase and DPNH (NADH) and TPNH (NADPH) cytochrome c reductases have been found to be associated with the microsome fraction, but he believes that the site of effective action of the other enzymes might be at the interface between the membranes and the matrix of the cytoplasm. He suggests that hexokinase might be activated at the membrane surface of the endoplasmic reticulum. Other cofactors such as glucose-1-6-diphosphate and adenosine monophosphate possibly also bind their appropriate enzymes to the E.R. membranes. (DPNH and TPNH are believed to act as coenzymes for glycolysis in the early stages of glucose oxidation.)

Siekewitz has also discussed certain biochemical aspects of the control of glucose metabolism.

First of all he points out that, if the concentration of glucose in the lumen of the endoplasmic reticulum is in equilibrium with the glucose concentration in the blood (this assumes at least temporary continuity between the lumen or cisternae of the E.R. and the exterior of the cell), there exists then a mechanism whereby the glucose level in the blood would definitely affect intracellular glucose equilibrium. Thus a reduced production of glucose from the diet would lead to reduction of glucose in the blood and this would cause a reduction of the amount of glucose in the fluid within the endoplasmic reticulum. The latter result would lead to increased phosphatase activity which would cause an increase in the breakdown of glycogen (e.g., in the liver cells) and a production of glucose, which would pass out from the endoplasmic reticulum into the bloodstream.

Siekewitz points out that hexokinase, phosphoglucomutase, and phosphorylase might have their activity enhanced if they were attached to the endoplasmic reticulum membranes. In the case of phosphorylase, active phosphorylase B has to undergo a conversion to active phosphorylase A before it can have any catalytic effect. An enzyme phosphorylase B-kinase carries out this conversion by phosphorylating the enzyme in the presence of ATP. Siekewitz suggests that this kinase may be part of the membrane of the endoplasmic reticulum and that the activating process for phosphorylase in the cell might consist of moving it out of the cytoplasm onto the site of the E.R. membranes. Siekewitz points out that it is possible that hormones control this type of movement of enzymes within the cell and its internal membranes. At this point one might remember the work quoted earlier by Brandes and the present author in which it was demonstrated that shifts of acid phosphatase activity took place from the Golgi apparatus to the nucleus in ventral-lobe prostate cells following castration and that acid phosphatase activity was restored to the Golgi apparatus following implantation of the male sex hormone. Also of interest are the further studies of Brandes in which he has demonstrated that in castrated animals there is a rearrangement of the membranes of the endoplasmic reticulum. These studies demonstrate a definite morphologicobiochemical effect on the part of the male sex hormone. Siekewitz suggests that the hormones concerned with carbohydrate metabolism do not act directly on the enzyme as such but bring it and the substrate and various cofactors together at a suitable surface and then complex them together there (see Fig. 58). Figures 59A, B, C, D, E, and F show the possible origin of an enzyme in the nucleus and its passage to the endoplasmic reticulum.

Other enzymes may be localized on the endoplasmic reticulum, e.g., 5 nucleotidase.

The nucleus, as will be described later, appears to be enclosed in a fold of the double membrane of the endoplasmic reticulum and not to have a membrane of its own. Since the E.R. spaces might be connected directly to the exterior of the cell, it is possible that glucose coming from outside the cell could be converted into glucose-6-phosphate in the E.R. lumen and thus be prevented from passing through the E.R. membrane. Thus it could pass along all

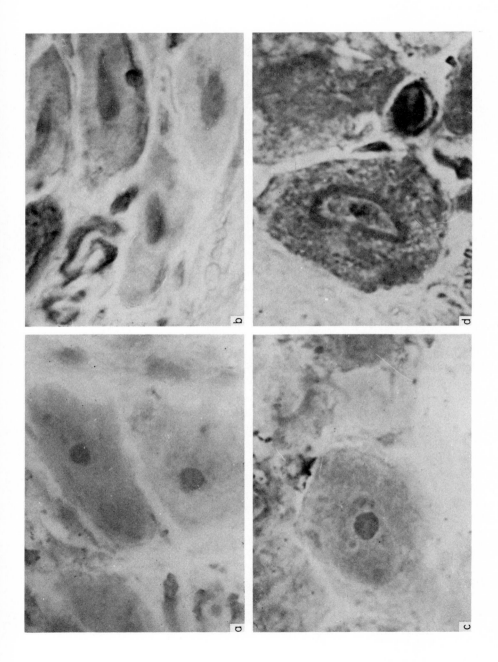

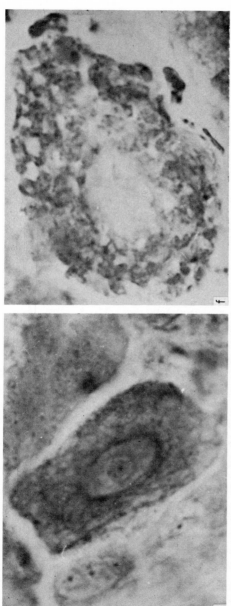

FIG. 59 Glucose-6-phosphatase in rat spinal ganglion neurons. (A) Note reaction restricted to nucleolus; (B) note reaction in nucleus, nucleolus still positive appears outside the nucleus and in contact with the cell membrane; (C) strong reaction in nucleolus, rest of nucleus and cytoplasm give moderate diffuse reaction; (D) moderate reaction in nucleus, stronger in cytoplasm, intense perinuclear reaction; (E) nucleolus and nucleus slight to negative reaction, reaction in cytoplasm appears to be associated with the endoplasmic reticulum; (F) completely negative nucleus and nucleolus. Cytoplasm filled with large negative vacuoles; cytoplasm between the vacuoles is strongly positive. These results suggest a cyclic metabolic activity in the spinal ganglion cells. It is possible that glucose-6-phosphatase is synthesized in the nucleolus and passed first to the nucleus and then to the cytoplasm where it becomes associated with the endoplasmic reticulum. The presence of large synthetic products in the cytoplasm suggest the participation of the glucose-6-phosphatase in some synthetic process. This product is not glycogen, but some of our protein and fat preparations show moderate sized droplets which could be identical with the vacuoles. (Preparations and photographs by H. B. Tewari, Dept. of Anatomy, Emory Univ.)

the ramifications of the E.R. canals and come in direct contact with the nuclear fold of the reticulum. If this membrane contains glucose-6-phosphatase (and a number of our histochemical studies suggest that it does, see Fig. 59), glucose could penetrate through into the nucleus without having to pass through the substance of cytoplasm at all. Histochemical preparations show that the glucose-6-phosphatase reaction in a particular histological section is not always positive for all nuclei in the section, and it is of interest that Siekewitz has pointed out that enzymes may be present or activated at the E.R. membrane only when the glucose concentration reaches a critical level.

It is of interest that histochemical study of the cells of many organs demonstrates the fact that many dephosphorylating enzymes as well as glucose-6-phosphatase show an association with nuclear membranes—perhaps here lies the mechanism whereby even low levels of glucose-6-phosphate could penetrate easily through into the interior of the nucleus. The endoplasmic reticulum could provide a pathway straight to the nucleus that would prevent glucose from coming into contact with or passing through the cytoplasm where it could be attacked by glycolytic enzymes. This may be the mechanism whereby a supply of glucose to the nucleus is ensured. Glucose could also be supplied to the nucleus from the cytoplasm by the process of glycolysis. In this case, glucose being formed from glycogen as glucose-6-phosphate would be dephosphorylated and pass through the E.R. membrane and into the lumen where possibly it would be rephosphorylated to prevent its passing back again and would thus move along the lumen to the nucleus. Hence the nucleus could get its glucose directly from the cytoplasm via the endoplasmic reticulum or directly from the outside (the latter, however, only if a direct connection really exists.)

In many cells the mitochondria can also be seen to be completely surrounded by endoplasmic reticulum. It is possible that the mitochondria themselves obtain their glycolytic fuel directly as a result of glucose passing through the E.R. membranes undergoing glycolysis there and the glycolytic products feeding directly to the mitochondria. Some authors have suggested that mitochondria are themselves no more than diverticula of the endoplasmic reticulum.

The control of the rate of different types of metabolism, particularly respiration, in the cell can depend on two main factors. First, a structural factor that brings into apposition the appropriate reactants and varies with the extent of the surfaces available for these processes to take place, and, second, it may also depend on some chemical feedback where an excessive production of one kind of compound inhibits its continued production or slows down further synthesis. Alternatively, the metabolism in any one particular direction may be affected by the absence of a limiting amount of some specific substance or compound in the reaction chain. In the case of glucose, there is a complex system of control of its production and use. For example glucose-6-phosphate activates glycogen synthetase, which is one of the enzymes concerned in the synthesis of glycogen, the effect of this being to stimulate the storage of excess glucose as glycogen. At the same time if glucose-6-phosphate accumulates in some tissues, it inhibits the enzyme hexokinase that would normally phosphorylate glucose, thus the net effect of the accumulation of glucose-6-phosphate is to cut down on the phosphorylation of more glucose. This process is known as "negative feedback."

Sir Hans Krebs* in a recent article entitled "Rate Limiting Factors in Cell Respiration," discussed the control of energy utilization and pointed out that, in unicellular organisms, energy can be obtained directly from oxidation if air is present but if air is not present then anaerobic fermentation takes its place and energy is obtained by this source. In this instance it is the supply of air or oxygen that regulates which of these mechanisms comes into use, and oxygen is, in fact, the "rate-limiting factor." In higher animals the ability to undergo fermentation or anaerobic oxidation is still present and can be particularly well demonstrated in muscle. Krebs pointed out that the chemical systems in the cell that are concerned with the function of regulation are all fairly simple reactions but that there is an elaborate interlocking of these reactions. By this he means that the individual reactants may take part not only in more than one reaction but in very many different processes. One of the difficulties in sorting out such a complex of activity is that not all the component reactions of this elaborate interlocking series are known and we are dealing with a heterogeneous system in which there are many varied membranes and different spatial arrangements of the various reactants. He also points out that regu-

* After whom the Krebs cycle is named.

lation is probably a matter of reaction velocities, some of which will be accelerated and some slowed down and the question that has to be decided is the degree to which any of these are rate limiting. Krebs illustrates this point by considering the amount of oxygen used by 4-ml sheep heart homogenate, which contained about 10% of tissue (see tabulation). Thus one can demonstrate that the addi-

Substrate added	O_2 (μmole) used by 4 ml suspension
None	17
Pyruvate	26.1
Succinate	32.6
1-Lactate	21.6
Citrate	20.8
α-Oxyglutarate	25.4
Fumarate	19.1
Acetate	21.1
Glycogen	15.4

tion of glycogen adds nothing to the oxygen uptake so the amount of glycogen present is not a limiting factor in this system. The same applies if glucose is added instead of glycogen. On the other hand, when acetate, pyruvate, or other intermediates of the tricarboxylic cycle are added, there is an appreciable increase in the rate of oxygen uptake. The fact that the oxygen uptake in this system can be increased if suitable substrates are added to the mixture demonstrates that the electron-transport system from DPNH (NADH) to oxygen is not being used to its full capacity; thus it cannot be the factor that is limiting the uptake of oxygen. Certain special substrates that are known to reduce either DPN (NAD) or flavoprotein seem to be able to increase oxygen consumption. Thus the limiting factor appears to be that the mechanism for the transport of hydrogen from DPN (NAD) or flavoprotein is not being used to its full capacity if these special substrates are not present. The limiting step, therefore, is really the first stage in the electron transport system. If it is found that a particular substrate increases the rate of respiration, this is due to the fact that the substrate reacts more readily or easily with DPN (NAD) or flavoprotein than any endogenous substrate already present. This is the reason why pyruvate or α-ketoglutarate or succinate are responsible for the stimulation of respiratory rate in the experiment quoted. However, even if we accept this we are still faced with the identification of the factor that decides the rate of reaction between substrate and DPN (NAD) or flavoprotein.

Studies with dinitrophenol, which uncouples oxidative phosphorylation from respiration, are of interest and help to throw light on this problem.

Perhaps we should first say a word or two about this action of dinitrophenol. The uncoupling of oxidative phosphorylation from respiration has been compared to putting a car into neutral gear and still leaving the engine running. If the respiratory activities are regarded as the engine and the phosphorylation as the process of making the car go, then dinitrophenol uncouples the engine from the transmission of the car, the engine continues to turn but the car does not move; in the cell the respiration goes on quite happily but no ATP is formed. Normally ATP is formed from ADP and inorganic phosphate, and thus the factor that limits the rate of oxygen consumption and oxidation of pyruvate in the normal system such as we have described is not really the amount of enzymes present but actually the level of either ADP or inorganic phosphate. In the experiments carried out by Krebs, inorganic phosphate was present in a fairly good concentration and further quantities added to the system did not stimulate respiration. Therefore it is almost certain that the limiting factor must be the amount of ADP which is available. These experiments demonstrate the type of chemical control that a single compound can exert on a whole chain of reactions. One should also remember that another mechanism, a structural one that controls the rate of respiration, is the rate at which glucose can enter the cell and this can be hormonally controlled, although the method of action of the hormone is not exactly known. It is of interest that one of the factors that probably affects the rate at which glucose can enter the cell (if, in fact, the endoplasmic reticulum is continuous with the outside of the cell) is the degree of complexity of the endoplasmic reticulum. If this structure develops many ramifications, as presumably it seems able to do in certain cells such as spermatocytes, as demonstrated by Fawcett, then the surface area available for glucose to enter into the cytoplasm of the cell and so be metabolized is enormously increased or, conversely, it may be decreased by a reduction in complexity of the reticulum.

To return to the subject of respiration and ATP formation, the reaction for this process can be given as follows:

$$C_6H_{12}O_6 \text{ (glucose)} + 6\,O_2 + 38\,ADP + 38\,H_3PO_4 \rightarrow 6\,CO_2 + 44\,H_2O + 38\,ATP$$

This is the general reaction and is a summary of all the complex intermediatary reactions that in the end simply produce carbon dioxide, water, and ATP. Since, as Slater and Houlsman have pointed out, cells contain only relatively small amounts of ADP, as soon as it is all converted into ATP the process of respiration will stop— ADP is thus the limiting factor. However, when the cell is stimulated to do work there is a breakdown of ATP according to the formula given by Slater and Houlsman,

$$38\,ATP + 38\,H_2O \rightarrow 38\,ADP + 38\,H_3PO_4 \rightarrow \text{work}$$

and, since ADP is now being re-formed, respiration can go on so long as there is some left to be resynthesized into ATP. The addition of further ADP will, of course, keep respiration going.

It is of interest that, if mitochondria that have been separated from the cell by differential centrifugation are permitted to stand for some hours at room temperature, the phenomenon of uncoupling (which can also be brought about by dinitrophenol) of the oxidative phosphorylation system from respiration takes place. This type of mitochondrial preparation is described as "aging" mitochondria, and it is tempting to speculate whether in senescing tissues there may not be a progressive uncoupling of respiration from oxidative phosphorylation or a progressive hydrolysis of ADP so that less and less of this becomes available for synthesis of ATP. It has been possible to isolate from mitochondria that have been aged in this way a heme compound that will actually produce this uncoupling reaction. It has been described and given the name "mitochrome" and is fundamentally a pigment. However, there appears to be a lipid component in this "mitochrome" heme-protein preparation that is the actual factor responsible for uncoupling, and the heme protein of the mitochrome is not, in fact, the uncoupling factor at all. Mitochrome is very similar in structure and form to cytochrome and is probably derived from it. Certain unsaturated fatty acids such as oleic acid are also found to be active as uncoupling agents, and the lipid isolated from the mitochrome particle also appears to contain an unsaturated fatty acid. The fact that an uncoupling agent can be produced *in vitro* this way is of considerable impor-

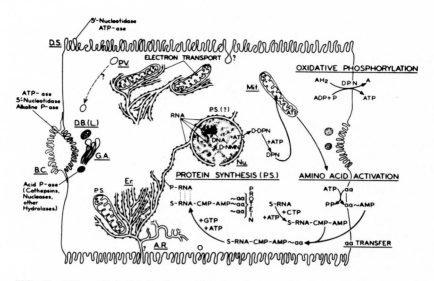

FIG. 60 Differential distribution of biochemical activities in the liver cell according to Novikoff. A, oxidized substrate; aa, amino acid; AH, reduced form of substrate; ADP, adenosine diphosphate; AMP, adenosine monophosphate; AR, agranular reticulum (smooth E.R.); ATP, adenosine triphosphate; ATPase, adenosinetriphosphatase; BC, bile canaliculus; CMP, cytosine monophosphate; D-DPN, desamido diphosphopyridine nucleotide; DPN, diphosphopyridine nucleotide; DNA, deoxyribonucleic acid; D-NMT, deamido nicotinic acid mononucleotide; DB, peribilary dense bodies; Er, ergastoplasm; GA, Golgi apparatus; GTP, guanosine triphosphate; L, lysosomes; Mit, mitochondria; Nu, nucleus; P, inorganic phosphate; P-ase, phosphatase; P-RNA, particle ribonucleic acid; PP, inorganic pyrophosphatase; PS, protein synthesis; PV, pinocytosis vacuoles; RNA, ribonucleic acid; S-RNA, soluble ribonucleic acid. (From Novikoff, *Am. J. Med.* **29,** 102, 1960.)

tance since it seems possible that the formation of such a substance in mitochondria might take place *in vivo* and may itself function as a controlling agent for respiration and oxidation. See Fig. 60, a diagram showing the relation of cell structures to function.

FAT METABOLISM

The general view of metabolism of fatty acids at present is that they are broken down beginning at the end of the chain where the carboxyl group is, and then the chain is progressively degraded as two carbon pieces are removed by a process of oxidation. This type of oxidation of fatty acids is known as β-oxidation and is so called because the fatty acid is attacked oxidatively at the β-carbon atom in the first instance. Among the products of this oxidation is the formation of acetyl coenzyme A and acyl coenzyme A. The latter can be subjected to further oxidation with the production of more acetyl coenzyme A (CoA). Although very little has been said about the mechanism of degradation of all these fatty acids, it is of interest that the enzymes which catalyze these reactions are all located in or on the mitochondria. The acetyl CoA can enter the Krebs cycle by condensing with oxaloacetic acid and the final oxidation thus follows the same path as the carbohydrates. The acetyl CoA formed from pyruvic acid (i.e., from carbohydrate breakdown) can be used to synthesize fats and likewise so can acetyl CoA produced as a result of protein metabolism. We see, therefore the reason why the Krebs tricarboxylic acid cycle has been spoken of as the meeting place of protein, fat, and carbohydrate metabolism—the final common path.

Where precisely in the mitochondria the enzymes responsible for β-oxidation of fatty acids are centered is not known for certain. It is very likely that they are more associated with the outer membrane of the mitochondrion than with the cristae so that the problem of penetration of the fatty acid through the membrane of the mitochondrion does not become so important. On the other hand, there is some evidence that pores occur in the mitochondrial membrane, and if this is the case it is possible for the long-chain fatty acids to pass into the interior of the mitochondria and to become subject to β-oxidation at this site by enzymes located in the cristae. Hoberman has shown that in the mitochondria, deuterium labeled DPNH (NADH), which has been reduced during the oxidation of fat in the organelles, is not available for reactions that take place in the outside cytoplasm. This suggests that the enzymes concerned with the oxidation are localized within the cristae. It is of interest that recent electron micrographs of the adrenal cortex have shown large areas of the surface where the mitochondrial membrane is incomplete, and the interior of the organelle is thus open to the penetration of the largest molecules and even particles of fat. Novikoff's scheme for the differential distribution of biochemical activities in liver cells is shown in Fig. 60.

Fat is metabolized largely in the way already described, but the synthesis of fat is also an important part of the activity of the cell. Fat cells have an important mechanical function to perform, and fat itself is a valuable reserve store of energy. Fats, fatty acids, and phosphorylated fats (phospholipids such as lecithin) are also important structural units of cell, mitochondrial, and other membranes, and their synthesis becomes an important cellular activity. Fatty acids are made up of long chains of carbon atoms, usually an even number, with a COOH group stuck at one end. Fat is formed from a fatty acid molecule by a combination between the latter and a molecule of alcohol such as glycerol. The link takes place through the COOH group which reacts with the OH of the alcohol to eliminate a molecule of water. Fatty acids may be short or long or intermediate chained, a typical short-chain fatty acid is acetic acid that has only two-carbon atoms, and a typical long-chain fatty acid is palmitic acid that has 16 carbon atoms.

In the synthesis of these long-chain fatty acids, the starting point appears to be acetic acid. This combines with CoA to form acetyl CoA. This condenses with CO_2 to form malonyl CoA (the coenzyme A ester of malonic acid). This latter compound then condenses with another molecule of acetyl CoA to form an intermediate substance that becomes reduced to form a four-carbon fatty acid that condenses with a molecule of malonyl CoA to give a six-carbon fatty acid and so the carbon chain is built up. Then, as described, combination of the appropriate fatty acid with an alcohol such as glycerol gives an ester known as a fat.

Phospholipids such as lecithin, which play a very important part in the structure of the cell membranes, also have to be synthesized by the cell. These compounds are not only esters of a fatty acid but are simultaneously esters of phosphoric acid. In this synthetic activity ATP plays an important part. It starts the ball rolling by phosphorylating one of the OH groups of glycerol. To the other two OH groups, CoA esters of palmitic acid are attached. At this point the phosphoric acid group that was put on in the first step drops off. The molecule then reacts with a compound known as cytidine diphosphocholine. This compound, which is really a coenzyme, drops the diphosphocholine part of its molecule nicely into the spot on the OH group that had been vacated by the first phosphoric group and so the lecithin is formed (see Fig. 61).

Other coenzymes are concerned with the production of the other steps in the synthesis but these will have to be studied in more specialized works of biochemistry. Dr. D. E. Green has pointed

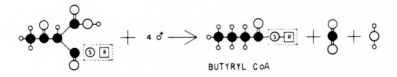

FIG. 61 Fatty acids build up from acetic acid units that are made reactive by combining first with coenzyme A to form acetyl-CoA (top) and then with carbon dioxide to form the CoA ester of malonic acid (second line). Malonyl-CoA and acetyl-CoA can condense (third line) into an intermediate compound. Further reactions not yet fully explained then reduce the intermediate to the CoA ester of a four-carbon fatty acid (fourth line). This, like acetyl-

out that one of the surprising things about the synthesis of fatty acids is that the synthesis stops at a limit of 16 carbons. It is rare to get 12- and 14-carbon chain fatty acids, and 18 or more carbon chains scarcely ever form. What tells the cell to break off the synthesis at 16 carbon atoms is certainly an intriguing problem.

Although one would expect that mitochondria would be the principal fatty acid synthesizers of the cell, the belief at the moment is that they are not the site of synthesis and that it takes place at the surface of the endoplasmic reticulum. However, in the production of fats from the fatty acids and especially in the case of lecithin synthesis where ATP is required, the mitochondria make a contribution to the synthesis because of their ability to produce the latter material. Mitochondria do play a direct role in fat metabolism but, this role, as mentioned earlier in this chapter, appears to be in fat degradation rather than in fat synthesis.

PROTEIN METABOLISM

Protein is composed of amino acids. Eight of these are nonessential and can be synthesized in the body and 12 are essential and have to be included in the diet (see Figs. 62–65). The breakdown of protein to its constituent amino acids takes place in the digestive tract; the amino acids are absorbed into the body and distributed to its cells. Further breakdown of protein by metabolizing their amino acids is therefore an intracellular process. Some amino acids are used for synthesizing protein, but those that are in excess of this requirement are degraded. Furthermore, when cells are starved they may break down their constituent proteins into amino acids and degrade these acids in order to supply energy. The amino acid is first deaminated, that is, the nitrogen in the form of an amino group (NH_2) is removed, and the rest of the molecule is oxidized. The nitrogen of the amino group is finally excreted as either ammonia, urea, or uric acid; in the case of a mammal it is usually as urea (the Dalmatian dog

---

CoA, can condense with a molecule of malonyl-CoA, ultimately giving the ester of the six-carbon fatty acid; the chain thus lengthens by successive steps. The molecules always join head to tail, the carboxyl head of the fatty acid joining the methyl tail of the malonyl-CoA. The letter R symbolizes the 82 atoms in coenzyme A other than sulfur. (This scheme is slightly modified from Lehninger, *Sci. Am.* **202**, 102, 1960.)

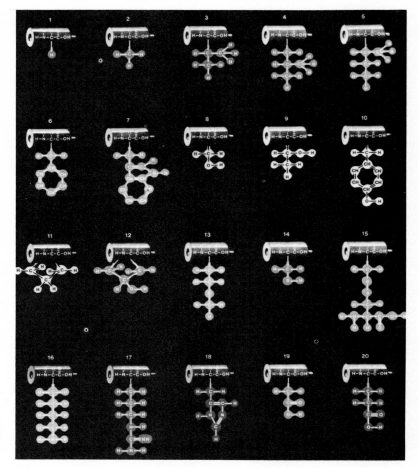

FIG. 62 Diagrammatic representation of the 20 x amino acids most commonly present in proteins. From a chemical point of view, an amino acid is an organic fatty acid in which the α-carbon atom has been substituted by an amino group. The fundamental property of amino acids is the presence in the molecule of an acid (—COOH) and a basic (NH$_2$) group. This means that amino acids can form bases with both acids and bases: in other words they are amphoteric. The structural part of the molecule which is common to all

$$\overset{R}{\underset{|}{}}$$

amino acids is H$_2$N—CH—COOH and this has been represented as a small cylinder with a short depression at one end and projection at the other—this serves to indicate the regions where the amino acids join together and underlines the fact that they may join up with each other in this way in any sequence. It is this possibility which permits the building up of a practically unlimited number of proteins. 1. Glycine, 2. alanine, 3. valine, 4. leucine, 5. isoleucine, 6. phenylalanine, 7. tryptophan, 8. serine, 9. threonine, 10. tyrosine, 11. proline, 12. hydroxyproline, 13. methionine, 14. cysteine, 15. cystine, 16. lysine, 17. arginine, 18. histidine, 19. aspartic acid, 20. glutamic acid. (From *Rassegna Med.* **42,** No. 4, 1965.)

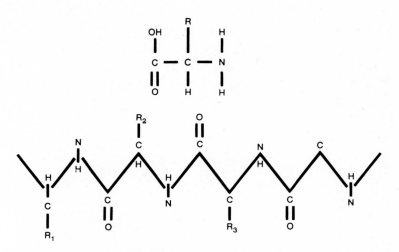

FIG. 63 How amino acids are linked together in a protein. An amino acid chain is shown schematically in the bottom half of the figure. The R's represent amino acid radicals—amino acid residues—and are different for each amino acid.

is an exception). The Krebs cycle (final common path) can convert some amino acids into carbohydrate. All the nonessential amino acids can be converted in this way into glycogen, but only a few of the essential amino acids can be used in some manner. Some of the essential amino acids can produce fatty metabolities; however this is not a reversible process so that fat cannot be used by the Krebs cycle to produce protein, whereas carbohydrate can. Little is known about the enzymes that effect the deamination of amino acids. It is not therefore possible to indicate where in the cell this process takes place. There are enzymes in the cell, known as transaminases, and a number of these have been localized in the mitochondria. Glutamic acid (one of the amino acids) can for example enter the Krebs cycle and be metabolized with production of energy. This is accomplished by the amino group being removed via transaminase, thus converting the glutamic acid into α-keto glutarate, which is one of the Kreb's cycle intermediates. Conversely transamination of α-keto glutarate will produce glutamic acid that is available for protein synthesis; similarly, the amino acid alanine can be deaminated to form pyruvate, which can enter the Krebs cycle, and

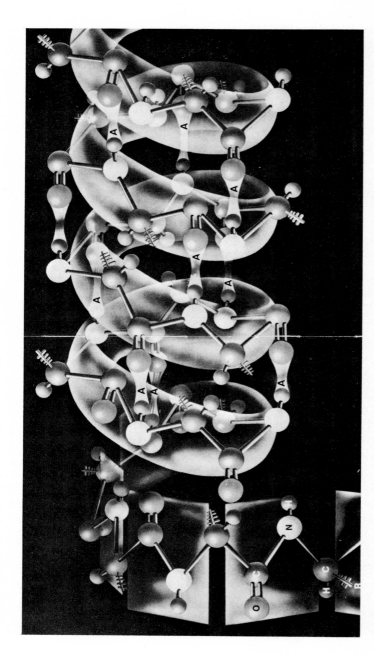

FIG. 64

pyruvic acid formed in the cycle can be transaminated to form alanine. In this way the mitochondria can form amino acids that in the endoplasmic reticulum can be synthesized into protein. Mitchondria play an active part in fat and carbohydrate metabolism and certainly are also concerned with the general metabolism of protein, though not specifically with its synthesis, which appears to be the function of the endoplasmic reticulum and its ribosomes. Mitochondria can, however, produce some protein themselves.

PLANT CELLS

So far we have considered only animal cells, and it is of interest that Hackett in the *International Review of Cytology* (**4**, 1954), has pointed out that in plant cells glycolysis involves the plastids, the soluble fraction of cells and possibly the nucleus as well as the mitochondrial enzymes. He points out that many enzymes that are involved in the Krebs cycle are not exclusive to the mitochondria and that hydrogen transfer is not confined to these organelles. Mitochondria, he says, react with the nucleus in the process of phosphorylation, they react with the chloroplasts in photosynthesis, and they react with the microsomes in protein synthesis in the endoplasmic reticulum. He believes that a close relation between the cell

---

FIG. 64 The process of transformation of an extended polypeptide chain into an alpha helical structure brought about by specific forces (hydrogen bonds, A) which are active between CO and NH groups. Alpha helices may be both dextro- and levo-rotary, the dextro forms seem to be more stable. Hydrogen bonds are themselves quite weak, but they are so numerous in molecules as large as proteins that they reinforce each other, giving marked stability to the protein molecule. The strength of our tendons and ligaments, the characteristics of our hair, etc., are all dependent on hydrogen bonds. Most of the specific properties and biological functions of each individual protein are ascribed to the differences between the side chains of each amino acid. Two or more protein chains may also join, assuming a crystalline structure defined as β-configuration. In this case the hydrogen bonds no longer determine the structure of the protein chain, as in the case of the alpha helix, but they bind the hydrogen atoms of one chain to the oxygen atoms of the adjacent chain. It is believed that during activity the polypeptide chains of muscle of other contractile fibers may pass from the alpha configuration to the beta configuration and vice versa. O, oxygen; C, carbon; N, nitrogen; H, hydrogen; R, side chain. (Figure and legend from *Rassegna Med.* **42**, No. 4, 1965.)

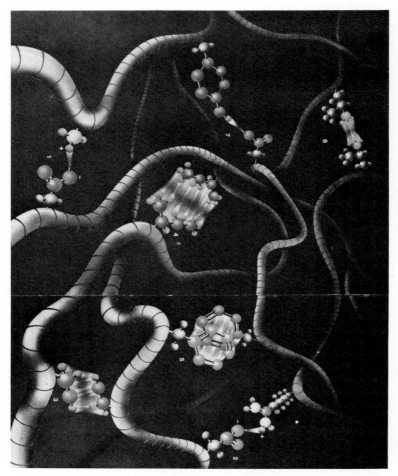

FIG. 65 A large number of proteins of fundamental biological importance
are "globular" in structure. This structure defined as "tertiary" is determined not only by the hydrogen bonds already mentioned in reference to the
helical structure but also by a series of forces which oblige the polypeptide
chain to bend in various ways causing the protein molecule to assume a more
or less interwoven and compact configuration. The forces are of the following
types: (a) Electrostatic; as between side groups of the type alanine–aspartic
acid, 1, and argenine–aspartic acid, 2; (b) van der Waals forces, serine–serine,
3, isoleucine–isoleucine, 4; (c) interactions between nonpolar side chains, due
to mutual repulsion of the solvent phenylalanine–phenylalanine, 5, leucine–
leucine, G; (d) hydrogen bonds, as occur for example between tyrosine
residues and carboxyl groups of side chains 7. (Figure and legend from
Rassegna Med. **42,** No. 4, 1965.)

membrane and the mitochondria may play an active part in the movement of substances into the cell or possibly in the growth of the cell wall of plants. Furthermore, he reminds us that the real unit is the cell itself, its various parts work as an integrated unit, and one should only break them down for the purpose of trying to analyze the various processes in which they participate.

FIVE
The
Golgi
apparatus

S all cytologists know, the Golgi apparatus was a structure first described in 1898 by the Italian neurologist
Golgi, in the nerve cells of the Barn Owl, although something like it appears to have been seen 30 years before
that. Since the discovery of this organelle it has been the
subject of very great controversy and in the last 50 years
there have been at least 2000 papers written about it.
The standard technique for differentiating the Golgi apparatus has been by the use of osmium or silver salts
after appropriate fixation of the cell, and the most characteristic form for the organelle that has been demonstrated is that of a network (see Fig. 66). In the case of
the nerve cell, this network extends pretty well through
most of the cytoplasm of the cell (see Fig. 80); in other
cells it is a small compact area situated to one side and
closely applied to the nucleus. The conception that the
Golgi apparatus existed as an osmiophilic or argentophilic
network was challenged by a number of workers. Parat
and his school believed that the Golgi apparatus really
consisted of a series of vacuoles that stained with neutral
red and an extension of this hypothesis, e.g., by John R.
Baker of Oxford, conceived of the network actually being
produced by the deposition of metals, osmium or silver, on
the periphery of the vacuoles so that eventually a network-

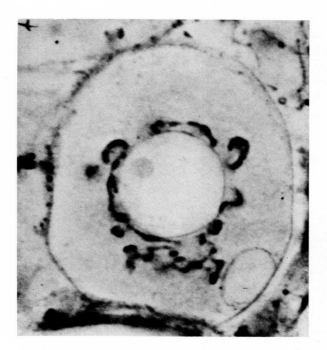

FIG. 66 The Golgi apparatus in a spinal ganglion cell. Notice its aggregation around the nuclear membrane. (Preparation and photograph by E. S. Horning, from "Cytology and Cell Physiology," Oxford Univ. Press, 1951.)

like structure was built up. Very complex functions were deduced for the Golgi apparatus, and it was believed to play a part in the formation of the acrosome of the sperm and to be concerned with yolk reproduction of the egg and secretion products in other cells, particularly those of the glandular cells. It is fairly certain that many of the structures described in various cells have been artifacts, in many cases critics of the Golgi apparatus as a cell organelle have themselves produced artifacts which resembled the Golgi apparatus and based their arguments on this artifact.

Dr. Pollister, in the *International Review of Cytology,* **6,** has discussed the structure of the Golgi apparatus and pointed out that many authors have described the apparatus as being lamellar in shape. He pictures it as being composed of irregularly circular, flattened, lamellae that surround one end of the nucleus. In various studies Pollister attempted to measure the thickness of these lamellae and found that they were in some cases thinner than 1 μ and in others around 0.25 or 0.20 μ. The apparatus is capable of considerable distortion and can be seen, in fact, distorted in the contracting smooth muscle fiber but appears to return to its normal shape once the force that is altering its shape is removed. For an excellent recent review of the Golgi apparatus see that by Beams and Kessel in the *International Review of Cytology*, **23**, 1968 (see Fig. 67).

Possibly one of the most interesting developments of the study

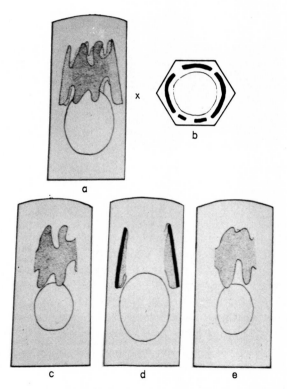

FIG. 67 The nature of the Golgi apparatus of the epithelial cells according to Pollister. (a) The whole collar in a vertical section of the epithelium, (b) a cross section of the cell taken at "X," (c to e) upper, middle, and lower focal planes, respectively. (From Pollister and Pollister, *Intern. Rev. Cytol.* **6,** 58, 1957.)

of the Golgi apparatus was its isolation from the epithelium of the epididymis of mouse and rat by Dalton and Felix in 1954. This was obtained by differential centrifugation from homogenates. It is also of interest that the Golgi apparatus appears to resist a good deal of damage to the cell. As long ago as 1925, Avel showed that, if living cells were ruptured by osmotic means, the Golgi apparatus was often one of the last parts of the cell to undergo disinte-

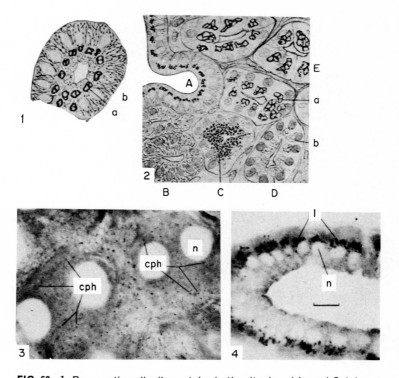

FIG. 68 1. Pancreatic cell, silver stain; both mitochondria and Golgi apparatus blackened by silver (from Cajal). 2. Submaxillary gland of young rabbit, silver stain; in one acinus, B, the mitochondria are blackened by silver, in another, (b) the zymogen granules, and in a third, (a) the Golgi apparatus (from Cajal). 3. Binuclear neurons of the coeliac ganglion of the rabbit colored with Sudan black, showing "cerephos" globules. 4. A frozen section of the intestinal epithelium of the mouse colored with Sudan black to show lipid globules in position of Golgi apparatus. (Figures and legends from Baker, "Mitochondria and Other Cellular Inclusions," *Soc. Exptl. Biol. Symp.,* Cambridge Univ. Press, 1957.)

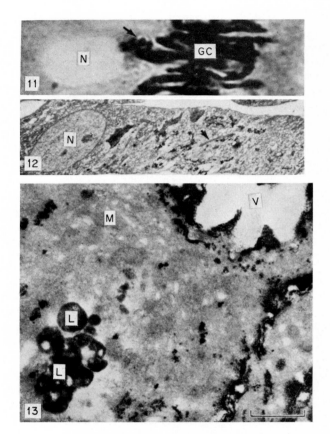

FIG. 69 Electron microscopy of the Golgi apparatus. 11. Golgi network in epithelial cell of mouse epididymus. Lipid droplets also visible (see arrow). 12. Low-power electron micrograph of a portion of an epithelial cell from the same block as 11. The Golgi network is also blackened. 13. Higher-power electron micrograph of section from same block as 11 and 12. A group of lipid droplets is seen at the lower left, and membranes and vesicles of the Golgi complex, all containing reduced osmic acid, are present at the upper right. (Figures and legends from Dalton and Felix, "Mitochondria and Other Cytoplasmic Inclusions." *Soc. Exptl. Biol. Symp.*, Cambridge Univ. Press, 1957.)

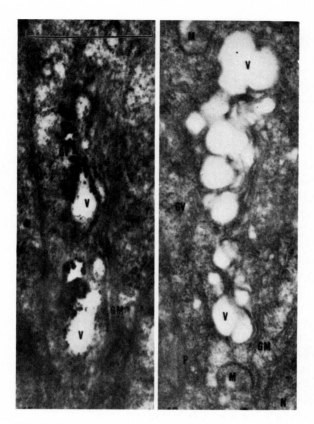

FIG. 70 Electron microscopy of Golgi apparatus. Left. Part of Golgi apparatus of mouse duodenal cell. Osmic acid has been reduced at the margins of the large vacuoles and within the small vesicles or granules but not in association with the membrane system. Right. Finer detail of Golgi complex of duodenal cell of mouse. (Legend and figure from Dalton and Felix, "Mitochondria and Other Cytoplasmic Inclusions," *Soc. Exptl. Biol. Symp.,* Cambridge Univ. Press, 1957.)

gration. The Golgi apparatus was extremely difficult to see in the living cell under the standard techniques of optical microscopy, even with the phase-contrast microscope (however, with better techniques, it was visible in germ cells).

The whole controversy about the Golgi apparatus appears to have been now reconciled by the studies with the electron microscope. The work responsible for this was done by Felix and Dalton as described in a series of papers published between 1953 and 1957 and also by Sjöstrand and by Rhodin. They showed that the Golgi apparatus consisted of a series of pairs of membranes adjacent to the nucleus which contained dilatations giving the appearance of vacuoles. (See Figs. 68–70, 74.) This structure thus explains many of the controversial results of the early workers. Sjöstrand, who contributed to the elucidation of this problem with his colleague Hansen, has stated that the membranes that constitute the Golgi apparatus form a system about 60 Å thick. These membranes are arranged in pairs. Along their edges there may be points of fusion followed by dilated areas (the vacuoles). The width of the space between the pairs of membranes (60 Å) is, apart from the presence of vacuoles, usually fairly constant (see Fig. 75). These pairs of membranes seem to be embedded in ground substance that is fairly homogeneous and has little structure, but in some cases it seems to contain fine granules or a fine reticulum. One of the characteristics of the Golgi apparatus that might otherwise have led it to be confused with the E.R. membranes is that the space between the latter, as previously described, measures 150 Å. Furthermore, ribonucleoprotein granules are attached to the outside of most of the E.R. membranes, but the membranes that constitute the Golgi apparatus are quite smooth and have no granules associated with them although there are reports of relatively large granules (400 Å across) being situated on or near the pairs of membranes in some cases. It is interesting that this characteristic structure that has been found for the Golgi apparatus is pretty well uniform in the cells of most species of animals and in cells belonging to a wide variety of organs; for instance, gland cells, nerve cells, muscle cells, and others.

The Golgi apparatus has been characterized throughout the literature as being that part of the cell in which the products of secretion are first recognizable microscopically, and Hirsch has even described in the living pancreatic cell small granules that were attached to the surface of the mitochondria becoming detached and moving through the cell cytoplasm toward the Golgi material.

The amount of information on the chemical nature of the Golgi apparatus is very large and very confused. We have just noted that the Golgi apparatus is composed of paired membranes that are dilated in parts to give the appearance of spheres, but this description of the Golgi apparatus does not satisfy all the various descriptions of the forms it takes in different cells when viewed by the light microscope. The relationship of this suggested ultrastructure to, for instance, the production of oögenesis and spermatogenesis has yet to be worked out. It is of interest here to note that Dr. H. B. Tewari has recently demonstrated that in the langur monkey, *Semnopithecus,* Golgi bodies pass from the follicular cells of the ovary into the developing cocyte where they appear to play a part in the elaboration of yolk. The passing of these bodies from one cell to another is the really fascinating part of this observation.

The electron microscope will probably, in due course, give us some information on the nature of some bodies described by Professor Hirsch as presubstances and what relationship they bear to a fully developed Golgi system. The significance of the difference in the physical nature of the apparatus in different cells will need to be explained, too. For instance, in most cells the specific gravity of the Golgi apparatus is less than that of the other cellular constituents as demonstrated by Beams and his colleagues by ultracentrifugation, but in uterine gland cells this is not so. In the cells of these organs it seems, in fact, to be a relatively rigid structure. Pollister has suggested that the Golgi apparatus could not be a fluid or even highly plastic solid, but that it has fundamentally a platelike form with enough elasticity to bend under pressure and straighten out again when the pressure was removed. Simpson, on the other hand, has claimed that it was always in a highly fluid condition. It is obvious that a good deal of the Golgi problem has to be reexamined and many light-microscope studies remade in view of the findings of the electron microscopists.

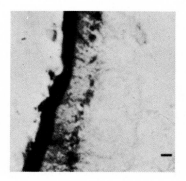

FIG. 71 Alkaline glycerophosphatase in epithelial cells of small intestine. In addition to reaction in brush border, there is a diffuse reaction in the distal part of the cytoplasm and a distinct reaction in the Golgi region. (Preparation and photograph by present author.)

The ability of the Golgi material to reduce osmium tetroxide, which it does very rapidly, is an indication that it may contain unsaturated lipid but then, of course, any reducing substance will reduce osmium tetroxide so this in itself is not a clear-cut histochemical test. However, the Golgi material has been stained on occasion with fat dyes, which suggests that it may contain fatty or lipoidal material. It has been claimed that the Golgi material breaks up after narcosis with chloroform. Monet has shown, that if the germ cells of *Helix* are treated with sodium bicarbonate, myelin figures are formed from the representatives of the Golgi material (dictyosomes). On the other hand, it has been claimed by Thomas that the dictyosomes of *Helix* are not Golgi material at all but are formed by the overimpregnation of mitochondria; this is the type of confusion that bedevils the subject of the Golgi apparatus. Ciaccio stained the Golgi apparatus and spermatids with his lipoid technique, and Boyle claims that part of the Golgi apparatus of the neurons of *Helix* are stained by Sudan IV, which is a fat stain. Baker carried out a series of detailed histochemical tests on the Golgi material and he found that the apparatus appeared to contain lecithin, cephalin, or sphingomyelin and that Windaus' test for cholesterol

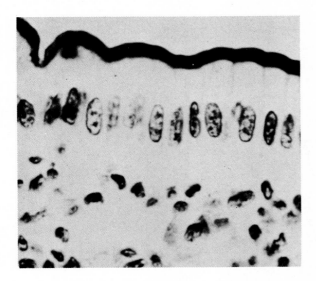

FIG. 72 Hexoestrol phosphatase in epithelial cell of small intestine. Strong reaction in brush border and nuclei but cytoplasm and Golgi apparatus completely negative. (Preparation and photograph by present author.)

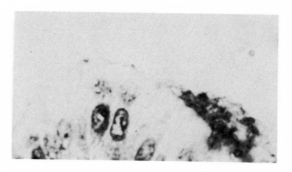

FIG. 73 Oestrone phosphatase in epithelial cell of small intestine. Strong reaction in nuclei and Golgi material, brush border negative. (Preparation and photograph by present author.)

and Schultz' test for cholesterol were negative. From the foregoing it seems reasonable to suggest that lipoidal material is present in the Golgi apparatus. This was confirmed by the studies of Schneider and Kuff in 1954, who isolated the Golgi material by differential contrifugation and found that it does contain a high concentration of phospholipid. As long ago as 1925, Nath suggested that the Golgi material contained protein. The same suggestion had been made by Bowen. Gatenby has expressed the opinion that the Golgi material is a combination of protein and lipoid. In fact some authors have found it possible to demonstrate the Golgi apparatus by fat dyes following a prior treatment of the tissue with proteolytic enzymes (pepsin and trypsin). They suggest that these removed the protein that was masking the lipoproptein complex and preventing the lipoid from reacting with the fat dye. Baker, using a variety of histochemical tests, found that the Golgi apparatus did not contain arginine or glutathione but that it gave a positive reaction with Millon's reagent, a positive xanthoproteic test and also a reaction for tryptophan. However, he found that the Golgi material was not colored more intensely than the cytoplasm so there was not a particular concentration of the amino acids that give these reactions in the Golgi apparatus. Protein has since been confirmed in the Golgi apparatus by a number of workers. This is not surprising in view of the electron microscopists findings that the membranes of the Golgi apparatus have the trilaminar structure found in other

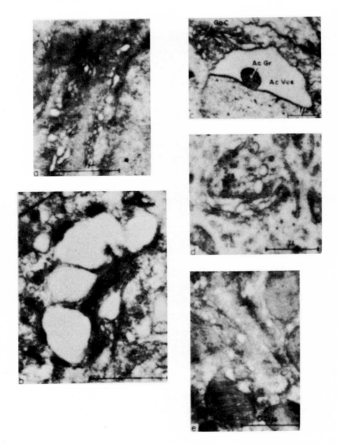

FIG. 74 Electron microscopy of the Golgi apparatus. Different types of apparatus in various cells. (a) Renal epithelium, mouse; (b) epithelial cell of epididymus, mouse; (c) spermatid of cat showing acroblast and acrosome; (d) juxtanuclear zone of Golgi apparatus in a human cancer cell; (e) portion of a renal epithelial cell of the mouse. The Golgi apparatus appears in the upper center. (From Pollister and Pollister, *Intern. Rev. Cytol.* **6,** 85, 1957.)

membranes in the cell that are known to be composed of lipid and protein layers.

The localization of vitamin C in the Golgi apparatus has a considerable literature and has been the subject of considerable controversy; for a detailed discussion on this subject, the reader is referred to a publication by the present author in *Protoplasma-*

tologia, "Vitamin C in the Animal Cell" and a comparable article in the same volume by Plaut on "Vitamin C in the Plant Cell." Evidence for the occurrence of vitamin C in the Golgi region has also been presented in the chapter on "Mitochondria and the Golgi Complex" in "Cytology and Cell Physiology" (3rd ed., Academic Press,

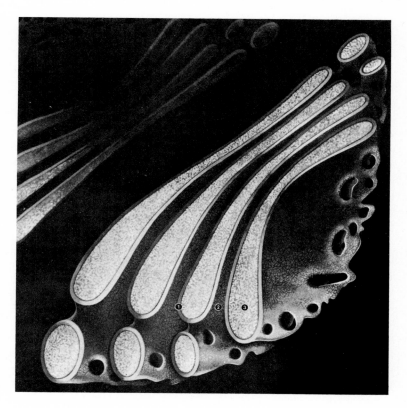

FIG. 75 Golgi's apparatus is made up of a system of simple membranes that enclose flattened spaces and form vesicles and vacuoles. This membrane system, lacking in RNA granules (smooth reticulum) often appears to be continuous with the endoplasmic reticulum. The function of Golgi's apparatus is still somewhat obscure. According to Palade, it seems to provide for the construction of new membranes, intended to substitute those destroyed through aging or damage. It furthermore acts as a "storehouse" for substances produced in the spaces between the layers of the rough reticulum. There may be more than one Golgi's apparatus in the same cell. 1. Sectioned membrane, 2. outer surface of the membrane, no ribosomes, 3. inner space of Golgi's apparatus. (From *Rassegna Med.* **45,** No. 3, 1968.)

1963). There appears to be a considerable identity in several types of cells between the Golgi preparations and preparations that are demonstrated by the application of the vitamin C reagent (acid silver nitrate). For example, in the neuron of the developing chick the Golgi preparation resembles very closely the result one gets from application of vitamin C reagent, and in the chick embryo liver a similar correlation can be seen. Diuresis causes changes in position of the Golgi material in rat kidney cells and this is comparable to the distribution of the vitamin C reaction. There is also a similarity of distrbution of Golgi material and vitamin C in fibroblast cells and in goblet cells in the rat colon. Furthermore the distribution of the vitamin C reaction in the ultracentrifuged adrenal, cortical, and medullary cells is identical with the position that is obtained with the Golgi stain. However, as has been mentioned before, the vitamin C reagent is extremely destructive of the cytoplasm of cells and it is very difficult when looking at its effects under the electron microscope to be dogmatic about localization of this material in the Golgi apparatus. Until further evidence is obtained, it is necessary to be conservative about the interpretation of vitamin C reactions in cells.

Some enzymes have been demonstrated to be present in the Golgi apparatus, for instance, it was first shown by the present author in 1943 that the columnar epithelial cells of the guinea-pig jejunum contained alkaline phosphatase in the Golgi region. The localization of this enzyme in the same region in the cells of the mantle edge that secretes the shell of the mollusk *Mytilus* was also figured. Various other authors have subsequently recorded the presence of this enzyme in the Golgi region. Deane and Dempsey, for instance, found that alkaline phosphatase was distributed in the Golgi region of the duodenal epithelial cells as granules or as a continuous reticulum (see Figs. 71–73). Activity was most intense in the Golgi region at the bases of the villi. Enzyme granules were found in the Golgi region in the cells of the kidney tubules, bile capillaries, and uterine epithelial cells of various mammals, and acid phosphatase has also been demonstrated in the Golgi region of the duodenal cells and in the uterine epithelial cells of pregnant cats and sows. Acid phosphatase has also been found in the Golgi region of the ventral lobe of the prostate by Brandes and Bourne, and details of the influence of the male sex hormone in maintaining its presence in this organelle has already been described. It is of inter-

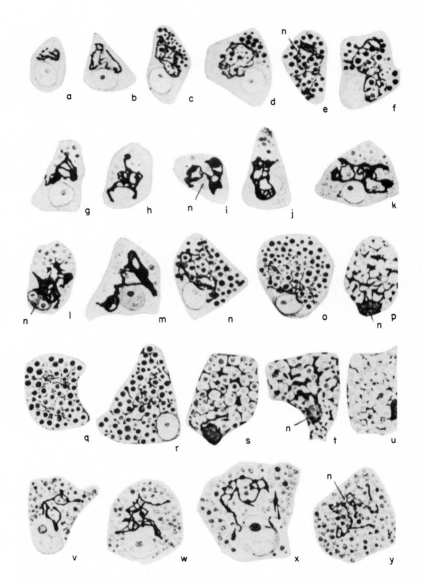

FIG. 76 Relationship of Golgi apparatus and secretion droplets in parotid gland of cat. (From Bowen, *Anat. Rec.* **32,** 151, 1926. By courtesy of the Wistar Inst. Press.)

est that the type of castration change described for the prostate could not be detected by biochemical studies of the cell, so that a biochemical investigation of the effects of castration or male sex hormones on acid phosphatase in the ventral lobe of the prostate would probably have shown no significant change in the level of acid phosphatase in the cells. Yet when this is examined by histochemical methods, it can be seen that the most fundamental changes have in fact taken place in the cell and a basic change in the locus of activity of an important enzyme has occurred.

Deane and Dempsey have also demonstrated that adenylic acid phosphatase (presumably 5-nucleotidase) has a similar distribution in the Golgi apparatus of liver cells as glycerophosphatase but it appears at a different pH. These authors suggest that all cells may

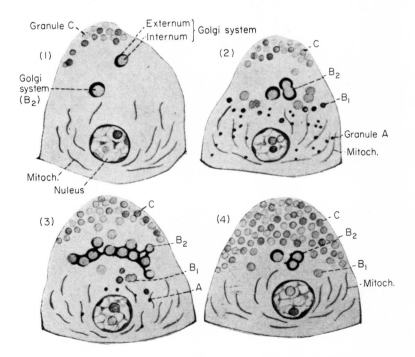

FIG. 77 Relationship of the secretion droplets to the Golgi material and mitochondria in the exocrine pancreas cell, according to Hirsch. Note small droplets associated with the mitochondria and moving toward the Golgi region. [From Hirsch, 1939 (Form und Stoffwechsel der Golgikörpern. Protoplasma Monographs, Berlin).]

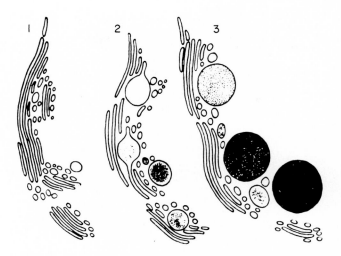

FIG. 78 Production of secretion droplets in Golgi lamellae (after Hirsch). 1. System is resting. 2. The system is building Golgi vacuoles and intermediate bodies. 3. Zymogen granules with membranes produced by Golgi lamellae.

have significant phosphatase activity in the Golgi zone at some pH and with some substrate. As far as the function of the phosphatase in the Golgi apparatus is concerned, nothing certain is known. Emmel has pointed out that the enzyme is present in the lumen of the intestine, kidney tubules, uterus, and bile cavities and suggests that it is being synthesized and excreted by the Golgi complex. This is of course a possibility. It may also be concerned, particularly in the absorptive cells of the gut, with the phosphorylation and dephosphorylation process concerned with the passage of some molecules through the cell membrane.

In the last few years the Golgi material that Dalton and Felix have been able to isolate from homogenates of epididymides has provided more specific information concerning its composition—at least in this organ. They found the isolated apparatus to be refringent and part of it to be extractable with 70% alcohol. Another part, however, was insoluble but this part was stainable with Sudan black, indicating the presence of some lipid; presumably it was a lipoprotein. Schneider and Kuff found with identical material that the pentose nucleic acid (RNA), the phospholipid, and the phos-

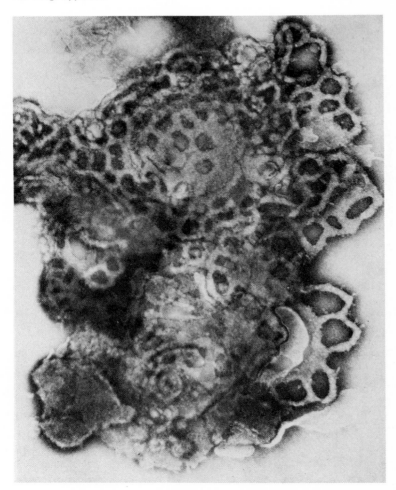

FIG. 79 Golgi apparatus from rat kidney. Removed from homogenate by differential centrifugation and negatively stained and examined under the electron microscope. (From H. W. Beams and R. G. Kessel, The Golgi Apparatus. Structure and Function, *Intern. Rev. Cytol.* **23,** 229, 1968.)

phatase concentration in the isolated Golgi fraction was greater than that in the whole tissue. Ascorbic acid, DNA, cytochrome oxidase, and DNAase were absent from the Golgi fraction of epididymal cells but the isolated Golgi material also gave a strong periodic-acid Schiff reaction that seems to have been due to a lipid component. A similar

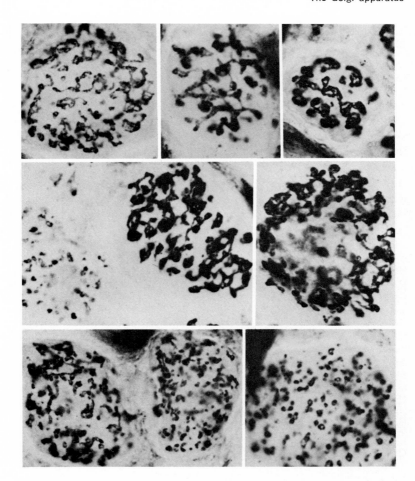

FIG. 80 Golgi apparatus of rat trigeminal ganglion cells, demonstrated by the thiamine pyrophosphatase technique. The material is arranged in the form of small-sized granules and vesicles of odd shapes in some cells, in others it is arranged as a complicated network. The stainability of the network varies from cell to cell. It is not uncommon to find cells with a network shaped Golgi apparatus lying next to cells appearing to contain only discrete Golgi vesicles. (Preparation and photograph by T. R. Shantha, *Acta Histochem.* **11,** 337, 1966.)

reaction has also been reported in intact cells from Leblond's laboratory (see Fig. 79).

In addition to the phosphatase which the Golgi apparatus of most cells has been shown to contain, other enzymes may occur. The present author's studies have demonstrated that oxidative enzymes are often present in the Golgi apparatus of Purkinje and other brain cells. It looks as though the composition of the Golgi apparatus from an enzymic point of view may differ from cell to cell and possibly from time to time, but further studies of this are required.

Novikoff and his colleagues found that the diphosphates of guanosine, uridine, and cytosine are hydrolyzed by the Golgi apparatus. This was reported not only in rat tissues but also in root tips of the corn plant by Dauwalder and his colleagues. Further

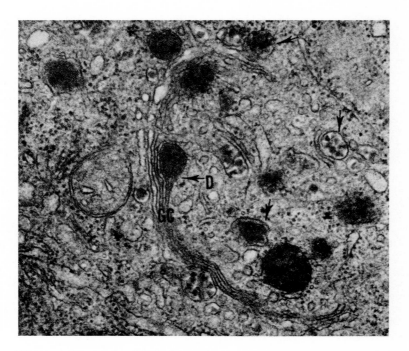

FIG. 81 Golgi complex (GC) in liver cell of young rainbow trout. Note multiple granules in dilated end of cisternae (D) and their similarity to nearby, isolated granules (unlabeled arrows). 53,000×. (From H. W. Beams and R. G. Kessel, The Golgi Apparatus. Structure and Function, *Intern. Rev. Cytol.* **23,** 229, 1968.)

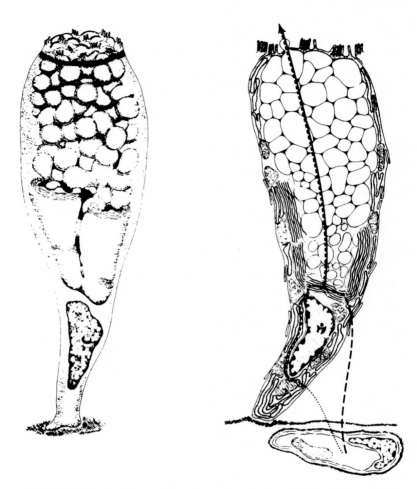

FIG. 82 Golgi apparatus and mucigen granules in goblet cell. (From H. W. Beams and R. G. Kessel, The Golgi Apparatus. Structure and Function, *Intern. Rev. Cytol.* **23,** 229, 1968.)

studies by Lazarus and Barden demonstrated that this diphosphatase activity was present in the tissues of a number of mammals. Another enzyme that has been found in the Golgi apparatus is thiamine pyrophosphatase, which hydrolyzes co-carboxylase, an essential component in many of the metabolic processes of the cell (see Fig. 80). The significance of this activity is not known. It was first discovered by Allen in the Golgi apparatus of cells of the epididymus and was subsequently found in the cells of other organs by Novikoff and his colleagues and by Shanthaveerappa and Bourne and other investigators. In fact, it seems possible that the thiamine pyrophosphatase reaction is a universal marker of at least part of the Golgi apparatus. It is of interest that when the Golgi vacuoles leave the Golgi saccules they lose their thiamine pyrophosphatase activity. Bradbury and Meek have studied the acroblast in the spermatids of the snail (the acroblast is a representative of the Golgi apparatus). The outer part of the acroblast was found to contain both phospho- and glyco-lipid, alkaline phosphatase, and thiamine pyrophosphatase; the inner part contained neutral mucopolysaccharide. They could not find ribonucleic acid and acid phosphatase in the acroblast. (See Fig. 81 for form of Golgi apparatus in acroblast.)

Esterase, glucuronidase, and peroxidase have also been found in the Golgi apparatus.

The relationship of the Golgi apparatus to secretion is shown in Figs. 76–78, and 82, and further details of the relationship of the Golgi apparatus to secretion are given in a later chapter.

SIX
The nucleus and the nucleic acids

THE general morphology of the nucleus is very well known and need not be considered in great detail here. Under the light microscope it appears to have little structure apart from the existence of refringent nucleoli. In certain kinds of cells, for example, eggs, the existence of nuclear membrane can be detected with the light microscope. Under phase-contrast microscopy more detailed structure can be seen in the nucleus itself, and most of this represents the chromatin that with appropriate fixing and staining appears as a network. The spaces between the network contain what might be described as a "sap" that varies in amount in different types of cells. The size of the nucleus is about 5–7 μ but might be much larger in some cells. The shape of the nucleus is usually spherical but may be altered in pathological conditions, and in some organs of senescent animals distorted nuclei can be seen. In old cells the nucleus may become pyknotic and stain excessively with basic dyes, and under certain physiological conditions the nuclear shape may change—a classic example of this is the shape of the nuclei of the silk gland cells of the silkworm. Here the tortuous shape of the nucleus, involving as it does a tremendous increase in the area of the membrane, must indicate a considerable degree of nucleocytoplasmic interplay. This subject will be referred to again in much greater detail. The shape of the nucleus may vary from a rounded spherical to even a branched shape as seen in the silk gland cells of the silkworm or it may even be segmented or have any kind of irregular shape as, for example, in the polymorphonuclear leucocytes of the blood. Most cells contain only one nucleus, but some, e.g., some liver cells, may contain two nuclei. In many cells such as those of the liver, the nucleus is more or less centrally placed in position; in epithelial cells it is usually situated at the base of the cell, i.e., near the blood vessels underlying the epithelium.

Nuclei may be "granular" in nature or have a "massive" or vesicular structure. In granular nuclei there are many small particles of chromatin scattered more or less evenly throughout the nucleus. "Massive" nuclei stain intensely, appear homogeneous, and do not have an obvious nuclear membrane. The most usual type of nucleus is the vesicular nucleus. This has a very obvious nuclear membrane with chromatin stuck along its inside and masses of chromatin in the body of the nucleus.

The nucleus is known to contain two types of nucleic acid, deoxyribonucleic acid and ribonucleic acid. It contains basic proteins and other proteins that include enzymes, phospholipids, various phosphate compounds, and a number of inorganic compounds.

There is now a considerable list of enzymes that have been found in the isolated nucleus. These include those that are related to the glycolytic cycle and to the oxidative cycle (although the latter are present in very much smaller amounts than in the cytoplasm). Enzymes concerned with the nucleotide metabolism are naturally present and enzymes that play a part in protein and fat metabolism also occur. A partial list of the enzymes found in the nuclei is given in Table I. In addition to these biochemical results which have been obtained, studies in the author's laboratory on the distribution of a variety of dephosphorylating enzymes which hydrolyze uridine, inosine, cytidine, and guanosine triphosphates and others that dephosphorylate DPN (NAD), TPN (NADP), and a wide range of other phosphate esters that are present in the nuclei of most cells.

Two enzymes in addition to those listed in the table are DNA polymerase that catalyzes the synthesis of nucleotide polymers to form DNA and RNA polymerase, which does the same thing for RNA.

Generally speaking these results indicate, as one might expect, the presence of most of the enzymes concerned with nucleotide

TABLE I
Nuclear enzymes

Enzymes in liver nuclei (unless otherwise stated), activity of total homogenate (%)	Enzymes in liver nuclei (unless otherwise stated), activity of total homogenate (%)
Glycolytic cycle	**Oxidative cycle**
Aldolase, 5–31	Aconitase, 8–14
Fructose-6-phosphatase, 2	Aconitase, 22.5 in cerebral cortex
Glucose-1-phosphatase (nil)	Isocitric dehydrogenase, 2–3
Glyceraldehyde dehydrogenase, (small amount)	Isocitric dehydrogenase, 14.1 in cerebral cortex
α-Glycerophosphate dehydrogenase, up to 22.3	Cytochrome c, 26.3
Hexose diphosphatase (nil)	Cytochrome oxidase, varies from a very low level to 16.5
Pyruvate oxidase, up to 15.4	DPN cytochrome c reductase, 12
Nucleotide metabolism	DPN synthesizing enzyme,[a] varies from 69–92
Adenosine deaminase, 2.08	Fumarase, up to 9.3
Adenosine deaminase, 6.0 in heart; other tissues (a trace)	α-Ketoglutaric oxidase (a trace)
Adenosine-3-phosphatase (trace)	Malic dehydrogenase (very small amount)
Adenosine-5-phosphatase, 48 (5-nucleotidase)	Succinic dehydrogenase,[b] 10.1
ATPase, 10–34	TPN cytochrome c reductase, up to 12
DNAase, 6.9	
DNAase, 37 in thymus	**Protein metabolism**
Nucleoside phosphorylase (present)	Amine oxidase, 20–30
Ribonuclease, 13.8–49	d-Amino acid oxidase (present)
	Arginase, 36
Fat metabolism	Cathespin, 31.2
	Glutaminase, 14.5
Octanoate oxidase, 2.8	Leucine amidase, 14.0
Oxaloacetate oxidase, up to 10.5	Glutamic dehydrogenase, 28

[a] The remarkable concentration of DPN synthesizing enzymes suggests that one of the functions of the nucleus is to supply DPN for the rest of the cell.

[b] It is of interest that histochemical reactions for succinic dehydrogenase in most cells show no trace of the reaction in the nuclei.

metabolism and with many of those concerned in glycolysis. However there is complete absence of a number of enzymes vital to the functioning of the Krebs cycle, but a surprising concentration of some others such as aconitase. The relatively large amount of cyto-chrome C is of interest.

In the absence of a Krebs cycle there is some doubt as to

whether the nucleus and in particular the nucleolus is able to synthesize ATP, although the nucleus contains plenty of ATPase. If the nucleolus synthesizes protein (evidence for this will be presented shortly) it needs ATP. Where does it get it from? The mitochondria have all the equipment for the production of ATP, and it has been.

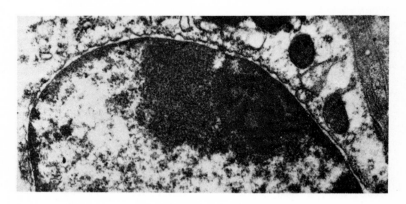

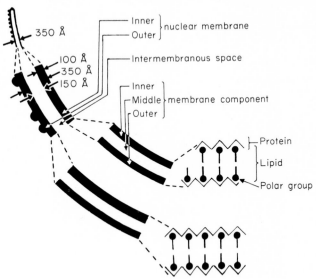

FIG. 83 Typical appearance, dimensions and molecular structure of the nucleus and nuclear membranes. Avian Brown Fat Cell. (From L. T. Threadgold, "The Ultrastructure of the Animal Cell," Pergamon Press, New York, 1967.)

noted that in tissue culture cells the nucleolus frequently moves in the nucleus to touch the nuclear membrane. Mitochondria have often been observed to touch the nuclear membrane at the same spot and at the same time as the nucleolus, and it has already been suggested that perhaps ATP is passed across the nuclear membrane at this time. It is of interest that Tewari and the present author have shown that in spinal ganglion cells there appears to be a cycle of activity between the nucleolus and the mitochondria. When the nucleolus moves to the side of the nucleus and touches the nuclear membrane, the mitochondria are clustered around the nucleus. The nucleolus then passes to the center of the nucleus and the mitochondria disperse through the cytoplasm. At the same time droplets of unidentified material make their appearance in the cytoplasm.

Before 1950, the electron microscope had given us very little help as far as the structure of the nuclear membrane was concerned, but during that year Callan and Tomlin were able to dissect out the large nuclei of the germinal vesicle of the oöcytes of amphibian eggs. These nuclei were ruptured by exerting a slight pressure on them, and the broken nuclear membranes so obtained were then examined under the electron microscope. By using this technique they found that the nuclear membrane actually contained pores. Callan and Tomlin then found that the nuclear membrane was composed of two sheets, an outer one in which the pores were present and an inner continuous sheet; these pores were approximately 400 Å in diameter. Pores of this type have also been found in the oöcytes of the starfish and in sea urchin eggs; the nuclear membranes of pancreatic cells have also been said to contain such pores. It has been claimed that neurons possesss as many as 10,000 pores in each nucleus. Although there has been some discussion about the reality of the existence of these pores, it is fairly certain now that they do occur more or less universally. In some EM photographs of the nuclear membrane, the pores can be seen to include both sheets of the nuclear membrane; in others it appears to penetrate only through one sheet (see Figs. 83–86).

The dimensions of each sheet of the nuclear membrane are of interest. Under the electron microscope the membrane can be seen to be composed of two osmiophilic layers with a clear layer in between. Each of the osmiophilic layers has a thickness of 70 to 80 Å, and the nonosmiophilic space between them ranges from

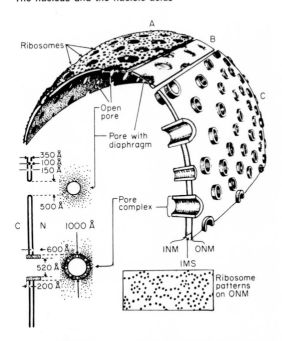

Ribosomes

Open pore

Pore with diaphragm

350 Å
100 Å
150 Å

500 Å

Pore complex

C N 1000 Å

600 Å

520 Å

200 Å

INM | ONM

IMS

Ribosome patterns on ONM

FIG. 84 Three dimensional representation of the nuclear membrane. (A) Open pores. (B) Pores with diaphragm. (C) Pore complexes. (From L. T. Threadgold, "The Ultrastructure of the Animal Cell," Pergamon Press, New York, 1967.) INM, inner nuclear membrane; ONM, outer nuclear membrane; IMS, intermembranous space.

100 to 150 A. It is of great interest that this is practically identical with the dimensions of the intracytoplasmic membranes (endoplasmic reticulum) and it has, in fact, been suggested by Watson that the nuclear membrane is not a separate membranous structure at all but is simply a part of the endoplasmic reticular membrane. If these can themselves be regarded as extensions of the cytoplasmic membrane then the nucleus is really enveloped in what is a deep and complex fold of the cell membrane.

The specific details of the nuclear membrane vary considerably in different types of cells but, generally speaking, the three-layered structure of the membrane and the presence of pores are characters that can be accepted for most nuclear membranes examined to date. The nucleus of nondividing cells has been referred to as being in a "resting" condition, and this is singularly inappropriate in view

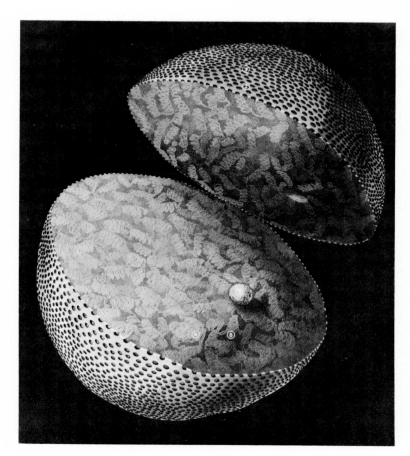

FIG. 85 The nucleus. Chromosomes extending through the matrix of the nucleus. The nuclear membrane represents the innermost limit of the cytoplasm; it is double (though only shown single in this diagram) and contains 1000 to 10,000 pores opening into the cytoplasm. The cell's set of deoxyribonucleic acid (DNA) is collected inside the nucleus in the form of chromatin. During the intermitotic phase it disperses to increase the surface of contact between DNA and the substances dissolved in the nuclear amorphous matrix. Before mitosis the chromatin condenses and forms chromosomes—a fixed number for each cell (46 in human cells). The nucleolus is clearly visible during the intermitotic stage; it is made up of granules whose morphology and composition are identical with those of the ribosomes of the rough endoplasmic reticulum. The nucleolus has no membrane. 1. Nuclear membrane with pores, 2. pores, 3. nucleolus, 4. amorphous matrix. (From *Rassegna Med.* **45,** No. 3, 1968.)

of the metabolic activity that this organelle must be carrying out almost continuously.

The relationship of the volume of the nucleus to that of the cytoplasm varies considerably in different types of cells, and, generally speaking, the volume of cytoplasm increases by comparison with the nuclear volume as the cells age.

Nuclei that are fixed by standard histological and cytological fixatives show a nuclear reticulum on which are scattered basophilic staining substances that have been described classically as chromatin, and the parts of the reticulum that do not show basophilic staining have been referred to as "linin." In and around this network it has been assumed that nuclear sap, which has been described as the "karyolymph" or "enchylema," exists.

The other obvious structure in the nucleus that can be seen is the nucleolus. There may be one or more of these and they vary considerably in shape. It is of interest that experimental studies of hybrids of the toad, *Xenopus laevis,* showed that there were some cells that contained nuclei without nucleoli and that this state resulted in the death of the cell [Elsdale *et al., Exptl. Cell. Res.* **14,** 642 (1958)]. The nucleolus was first described as long ago as 1781 by Fontana, and from then on it seems to have been pretty generally recognized as a typical cell structure. Nucleoli under the light microscope usually have the appearance of rounded homogeneous structures always within the nucleus, but Tewari and the present author have seen them attached to the outside of the nuclear membrane in spinal ganglion cells. There appear to be two structural phases in the nucleolus, one is in the form of a series of highly coiled strands and the other a structureless phase, presumably some sort of a nucleolar sap. This nucleolar sap is called the "pars amorpha," and the coiled strand is the "nucleolonema" described by Estable of Montevideo and his co-workers. The nucleolus contains a good deal of ribonucleic acid and there is also some evidence that it contains deoxyribonucleic acid as well. Caspersson has claimed that it is rich in diamino acids, and the presence of protein masked phospholipids has also been described; in the spinal ganglion cells the nucleolus stains intensely with mitochondrial techniques, but further details of the chemical nature of the nucleolus will be given later on. The nucleolus varies in shape and size according to the activity of the cell, during the period of anabolism it is said to become hypertrophied and during the catabolic stage

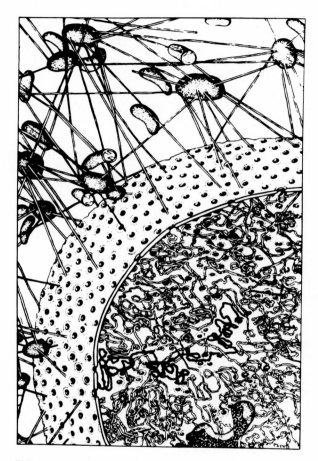

FIG. 86 Interpretation of nuclear ultrastructure in honey bee embryonic cells. One chromatin fiber, possibly equivalent to a single chromosome, has been defined in black and terminates at the inside margin of an annulus. The nuclear membrane also serves for structural orientation of extranuclear particles which are attached to the membrane by cytoplasmic fibers. (From E. J. DuPraw, "Cell and Molecular Biology," Academic Press, New York, 1968.)

it is said to become reduced in size. The nucleolus shows changes with pH; a number of drugs cause it to change in size, and it is also affected by ionizing radiations. Although the nucleolus is an area in the nucleus where there is a great concentration of special substances, the electron microscope up to date has given no evidence that it is surrounded by a limiting membrane. In the interphase nucleus, the nucleolus retains its characteristic spherical shape, and, provided the cell is not subjected to any particular type of stress, the size and general appearance of the nucleolus remain unchanged. Electron-microscope studies have demonstrated that there is in fact a filamentous element in the nucleolus that varies between 90 and 180 mμ in width and is actually composed of a number of fibrils varying 80–100 Å in width. In addition to the filaments many granules are present that are similar in appearance to the ribosomes of the cytoplasm and probably, like them, composed of ribonucleoproteins. Another component, the nucleolus, according to Bernhard, is a more or less homogeneous area in which are scattered a number of granules and a number of fine threadlike structures. The filamentous part seems almost certain to be Estable's "nucleolonema," whereas the homogeneous area is equivalent to his "pars amorpha." In nerve cells the nucleolonema is in the form of a network with affinity for silver salts. Using improved embedding media and better fixation, this network in the nerve-cell nucleolus was seen to be composed of a number of fibrils (50 Å in diameter and about 350 Å long) that were packed together. Some of these fibrils according to Bernhard ("The Nucleus," Vol. 3, Academic Press, New York, 1968) show a "double stranded structure."

Associated with this network and lying in its interstices are a number of granules about 150 Å in diameter. It was also found that the nucleolus was in fact surrounded by a concentration of chromatin (known as nucleolar-associated chromatin) that actually forms the boundary between this organelle and the rest of the nucleus. During the prophase stage of cell division according to Estable and Sotelo the nucleolus demonstrates a relationship to a specific chromosome as it develops.

However, detailed studies with the electron microscope have shown that this is not so. During the prophase the nucleolar material, ribosomes and fibrils, was found to be scattered through the nucleus. In Chinese hamster cells, however, the nucleolar organization may be retained attached to the chromosomes or free in the cytoplasm. At telophase, just before the nuclei re-form, small basophilic bodies appear between the chromosomes that form up to produce the nucleolus again. Although there is some evidence that the nucleolus is in a semifluid state, it has a considerable density and can be centrifuged to one end of the nucleus and more or less pure nucleolar material can be obtained by differential centrifugation. The studies with x-ray absorption and ultraviolet absorption suggest that the nucleolus is a semisolid body and that it contains proteins that are in a state of considerable hydration. It is of interest that in cells that engage in active synthetic activity the nucleoli become hypertrophied, and cells that are not actively synthetic have little or no nucleoli. The classic example of this can be found in the case of muscle. Embryonic cells that are forming muscle fibers have very well defined nucleoli, presumably because they are concerned in synthesizing muscle protein, whereas the nuclei in muscle fibers of adult animals show very few nucleoli and those that are present are extremely small. Starvation also causes a great reduction in size of nucleoli but they increase in size very substantially on refeeding, suggesting again that they are playing a role in protein synthesis. In embryonic tissue, nucleoli always become most obvious at the time when the tissue is differentiating, presumably at the time when specially differentiated protein for building up a specific organ is being made. Studies using DNAase have shown that the nucleolar-associated chromatin is composed of DNA. Likewise, using RNAase, it has been demonstrated that the granules and fibrils within the nucleolus contain RNA.

Since DNA acts as a template for the synthesis of RNA, this morphological arrangement is particularly significant. The amorphous matrix of the nucleolus is digested by pepsin and is therefore protein in nature. There is strong evidence based on autoradiographic studies that nucleolar RNA is discharged into the cytoplasm; there is also cytological evidence that this is so. This has been recorded both for fixed and for fresh tissues, and Duryee (1950) mentions that it takes place in the living oöcytes of amphibia. Tewari and the present author have seen it in spinal ganglion cells (see Figs. 90A, B, C, and D). It therefore appears quite possible that this may occur and will be referred to again later.

The rest of the nucleus is composed of nucleoplasm that does not appear to have a highly organized structure as demonstrated under the electron microscope but rather seems to be built up of irregularly distributed particles of various sizes although their average diameter according to Sjöstrand is 170 Å. The range of variation being 150–190 Å.

CHROMOSOMES

Chromosomes appear to organize themselves out of the reticulum of the nucleus during the process of mitosis, and during the preliminary stage of the prophase they become shorter and detach themselves from any reticular material left in the nucleus. Subsequently, with the disappearance of the nuclear membrane, the nuclear sap diffuses into and becomes mixed with the cytoplasmic materials. At this stage production of a spindle begins and the chromosomes lie free in the cytoplasm and then become organized in relation to the spindle that forms around them. Chromosomes are bodies that have been known for a very iong time, and even those of the insects have been known since the middle of the 19th century.

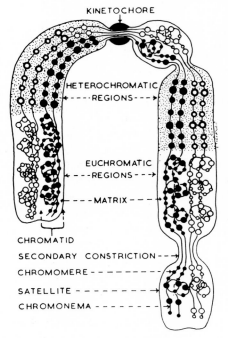

FIG. 87 Structure of the chromosomes. (From L. T. Threadgold, "The Ultrastructure of the Animal Cell," Pergamon Press, New York, 1967.)

Insect salivary-gland chromosomes are extremely large and show a very complex banding, which appears to be related to the location of the genes (see Fig. 87). These bands appear to contain DNA and can be demonstrated by a number of staining methods.

The persistence of chromosomes in the resting nucleus has been a problem to cytologsts for many years, and studies with ultra-violet absorption on the nuclei of living cells have demonstrated that at least the DNA of the chromosomes is still present at this stage (see below). If resting nuclei are broken up and ultracentri-fuged, fine filaments can be obtained from them, and it appears that these are probably greatly extended chromosomes. Certainly there is some evidence that typical chromosome structures are present in them. The bands that are shown to be present on chromosomes are known as euchromatic bands and heterochromatic bands (see Fig. 91). The former were said to be composed of DNA associated with histones, and the heterochromatin appears to contain both DNA and RNA. Curious types of chromosomes known as lamp-brush chromosomes have been described in the eggs of amphibia, fish, reptiles, and birds and they, like the other types of chromosomes, are banded but also have a number of loops that project from the surface and extend into the nuclear sap in which the chromosomes lie. These chromosomes are extremely long and very thin, and they develop their particular shape just about the time when yolk is being synthesized actively by the egg. At this time, too, the nucleus appears to contain quite a number of nucleoli, and it is thought that these are produced by specialized regions in the chromosomes. When oögenesis is complete the chromosomes shrink and become very much thicker and the number of nucleoli becomes reduced. The main axis of the lampbrush chromosome contains a considerable amount of DNA, and the loops that project into the cytoplasm seem to be composed largely of RNA. It has been demonstrated by Ficq that precursors of RNA and proteins that have been labeled with radioactive atoms are localized more in the nucleus than in the cytoplasm and that the incorporation of, for example, adenine into RNA seems to be localized almost exclusively in the loops of the lampbrush chromosomes. It seems pretty certain that the loops of these chromosomes represent highly active areas that are specifically concerned with the synthesis of protein.

Studies on the internal structure of chromosomes show that they are composed of one to four filaments known as "chromo-

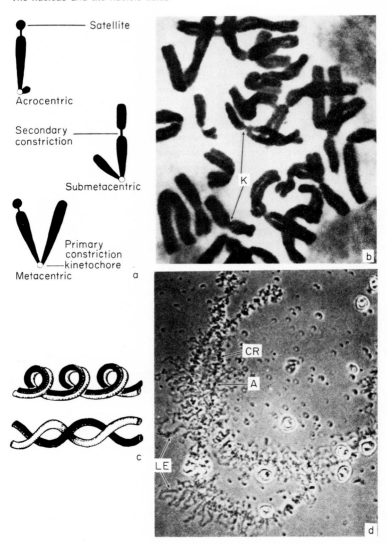

FIG. 88 Form of chromosomes. (a) Three types of chromosomes based on position of kinetochore. (b) Prometaphase chromosomes showing bipartite structure. (c) Types of chromosomes coiling: above, paranemic; below, plectonemic. (d) Lampbrush chromosomes with lateral expansions. (From L. T. Threadgold, "The Ultrastructure of the Animal Cell," Pergamon Press, New York, 1967.

nema." These are coiled, but it is not known whether they are wound round each other (plectonemic winding) or are simply coiled along side each other (paranemic winding). Some parts of the chromosome have regions known as "heterochromatic regions"—here RNA and DNA are believed to be concentrated. Other regions of the chromosome are known as "euchromatic regions" and are believed to contain DNA and histone.

Each chromosome has a region of primary constriction that might almost be described as an articulation, which is known as the "kinetochore." The position of the kinetochore decides the form that the chromosome takes at metaphase. If the kinetochore is terminal or almost so, then the chromosome is virtually without articulation—this is called an "acrocentric" chromosome. If the kinetochore is more centrally situated but still not in the center so that the chromosome has two unequal arms, it is known as a "metacentric" chromosome. In addition to the kinetochore or "primary" constriction

FIG. 89 The chromosomes of the Orangutan *(Pongo pygmaeus)*. Right bottom, appearance of chromosome in a squash. Above, chromosomes arranged in pairs. (Preparation and photograph by C. Hill.)

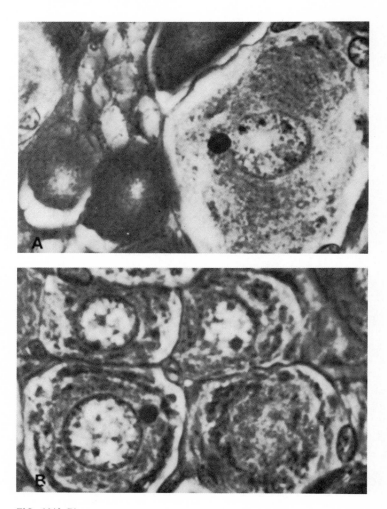

FIG. 90(A, B)

FIG. 90 (A–D) Spinal ganglion cell of rat, mercury bromphenol blue reaction for proteins. Showing passage of nucleolus out of nucleus to periphery of cytoplasm. (Preparation and photographs by H. B. Tewari, Dept. of Anatomy, Emory Univ.)

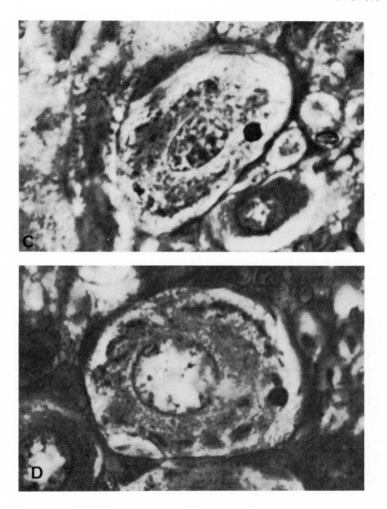

FIG. 90(C,D)

there are other less deep, secondary constrictions. Some chromosomes also have a rounded body known as a "satellite" attached
to them by a filament of chromatin (see Figs. 87–89).

Close to the nucleus is a structure known as a centriole, which
appears to play an important role in cell division. It is of interest
that under the electron microscope the centriole has many points
of similarity with the basal bodies found in ciliated epithelial cells.

The basal bodies are fundamentally cylindrical structures and so are the centrioles, the two bodies are also made up fundamentally of nine parallel tubules. Bernard and de Harven have indicated that each of the nine tubules are themselves made up of three smaller tubules that are in a turbinelike "disposition." According to de Harven ("The Nucleus," Academic Press, New York, 1968), the description "centriole" should be "restricted to cylindrical structures showing a ninefold rotational symmetry" (see Figs. 92–94).

Associated with the centriole are pericentriolar structures. These appear as small dense bodies about 70 μ across and have been described as "massules." Each centriole is said to be surrounded by two groups each composed of nine massules, and the latter are said to be connected with the centriole itself by bridges of an electron-dense material. These pericentriolar bodies are also known as "satellites" and, although all authors confirm their existence, they do not agree as to their number.

Within the centriole are formed the "intracentriolar structures." The tubules themselves show some diffuse osmiophilic material. located in their interiors. Various photographic techniques have demonstrated what has been described a "hub with nine spokes" occupying the center of the centriole but while some authors accept these, de Harven (see reference above) does not, believing them to be a photographic artifact. Little is known of the chemical nature of centrioles, although there is some tentative evidence that they may contain both RNA and DNA. Centrioles appear to duplicate themselves by growing out a daughter cylindrical structure at right

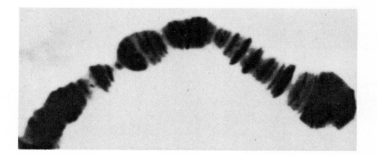

FIG. 91 Portion of a salivary gland chromosome in *Chironomus*. The dark bands represent localization of DNA. (From White, "Cytology and Cell Physiology," 2nd ed., Oxford Univ. Press, London and New York, 1951.)

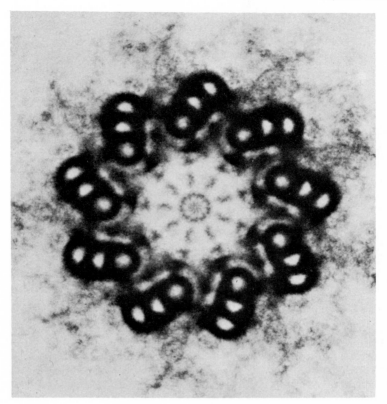

FIG. 92 Electron micrograph of a centriole. (From E. de Harven, The Centriole and Mitotic Spindle, *in* "The Nucleus," Academic Press, New York, 1968.)

angles to the long axis of the centriolar tube. Some authors have described granules that are adjacent to the main centriole and from which the daughter centriole is organized. Whether the method of formation of a new centriole is different in the cells of different organisms or whether there is a common form of centriolar duplication it is not possible to say at this time (see Figs. 92–94).

SPINDLE FIBERS

The fibers that make up the mitotic spindle appear to be canaliculi and they are approximately 20 mμ (200 Å) in diameter, that is,

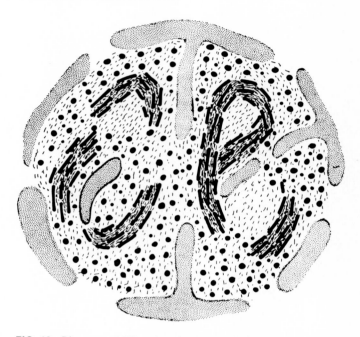

FIG. 93 Diagram of Ultrastructural organization of a normal nucleolus in vertebrate cells. Associated perinucleolar and intranucleolar chromatin is present. The RNA is linked to fibrillar and granular components. Both are embedded in a diffuse protein matrix. (From W. Bernhard and Nicole Granboulan, The Nucleolus in Vertebrate Cells, *in* "The Nucleus," Academic Press, New York, 1968.)

about the same diameter as the small tubules of the centriole. It is of interest that microtubules found in the cytoplasm of the interkinetic cell have the same size and form as the spindle fibers and some authors believe that the spindle is produced by the reorientation of pre-existing spindle fibers. It seems that the walls of the microtubules themselves have a microstructure, being made up of "protofibrils." The tubules themselves have shown no evidence of branching, during mitosis they attach themselves to the chromosomes in the region of the kinetochore, while some fibers extend from the centrioles to the chromosomes others extend from the centrioles to the other fibers. In each mitotic spindle there appear to be about 500 fibers.

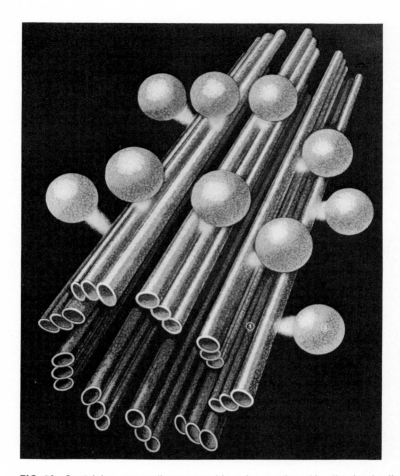

FIG. 94 Centrioles, generally arranged in pairs, are found in all animal cells. Each cell inherits a pair and synthesizes another. During the anaphase the chromosomes give origin to two new nuclei by migrating to the centrioles' poles. The centrioles form cylinders that are 3–5 μ long; their walls are made of thin parallel tubular structures, nine in number, each consisting, in turn, of three subunits. The satellites, or pericentriolar bodies, are connected to the centriolar tubules; little is known of their significance. Structure substantially similar to centrioles are found at the base of cilia and flagella. 1. Centriolar tubule made up of three subunits, 2. filament joining the satellite to the tubule, 3. satellite or pericentriolar body. (From *Rassegna Med.* **45,** No. 3, 1968.)

DNA, RNA, AND GENES

The discussion of chromosomes leads us on to the concept of the gene. It appears to be generally accepted that genes can be described chemically as nucleoproteins. The difficulty, however, is to find some way of explaining why these various genes are able to exert, either by chemical alterations, or physical structure, their different activities. It has been suggested that the difference in the proteins associated with the DNA in each of the genes is the important factor. At least this was originally thought to be the case, but now the sequence of nucleotides in the nucleic acid or in the spatial relationship of the nucleotides to each other is accepted as the key to differential action; the length of the nucleotide chain that makes up the DNA may also be important. Stern has suggested that one of the analogies concerning the genes was that the relationship of the different organization of DNA to production of different types

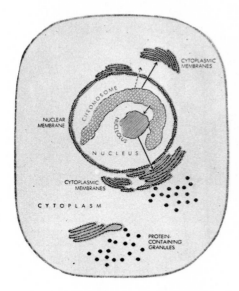

FIG. 95 Distribution of DNA and RNA, two substances that determine the form and function of cells. DNA (broken hatching) and RNA (solid hatching) are both found in chromosomes. RNA migrates (solid arrows) often via the nucleolus and nuclear membrane into the cytoplasmic membranes, where it presides over protein synthesis. DNA leaves the nucleus (broken arrow) only in certain cells. (Figures and legend from Gay, Nuclear Control of the Cell, *Sci. Am.*, January 1960.)

of genes was "the analogy of the sound track traced in a plastic matrix by a recording stylus, the modulated grooves would be the counterpart of genic modulations engraved on a chemically uniform nuclear protein matrix." According to this point of view the various genes would represent isomers or stereoisomers of each other. A structural gene is also known as a cistron. Before we consider genetic information further, we should study in more detail the nature of DNA and RNA.

We have mentioned RNA and DNA frequently, and we should now consider these substances in a little more detail. They are both nucleic acids and constituents, not only of the nucleus but as we have seen of the nucleolus as well and both RNA and DNA are also present in the cytoplasm, the former in association with the endoplasmic reticulum as ribonucleoprotein granules and occurring scattered through the rest of the cytoplasm (see Figs. 95–97).

The discovery and identification of nucleic acids began with the work of Miescher in the last half of the 19th century. Nucleic acids are found in all living cells, not only in animals but also in plants, and are associated with proteins to form nucleoproteins.

Nucleic acids are actually polymers, being composed of a very large number of nucleotides that represent the monomeric part of the polymer. They all have a basic chemical structure being composed first of a base that can be either a purine or a pyrimidine,

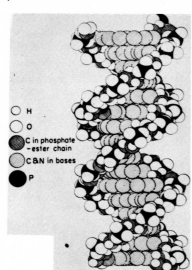

H
O
C in phosphate
–ester chain
C & N in bases
P

FIG. 96 Model of molecular structure of DNA, according to Feughelman *et al.*, 1955. (From Brachet, "Biochemical Cytology," Academic Press, New York, 1957.)

second a sugar, and third a phosphate group. There are two types of sugar present in nucleic acids, d-ribose and deoxyribose; nucleic acids have been divided into two main groups according to which type of sugar they possess. Those containing the ribose

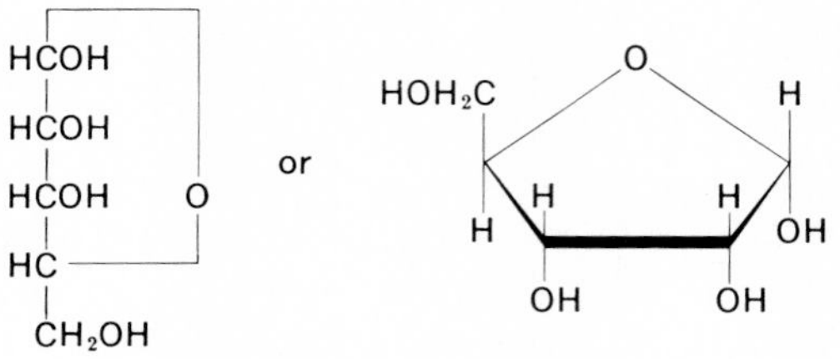

d-Ribose (α-d-ribofuranose)

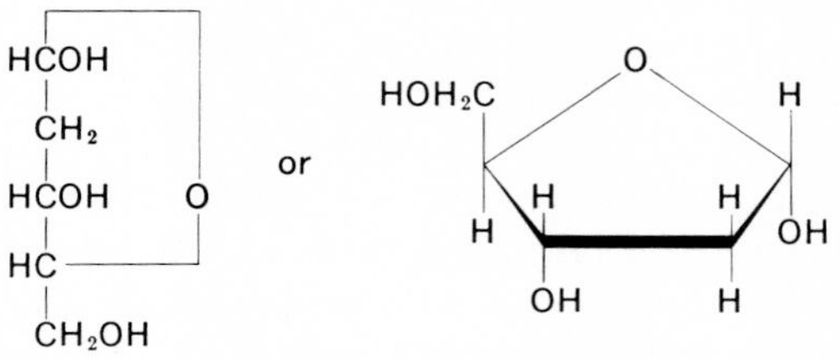

d-2-Deoxyribose (α-d-2-deoxyribofuranose)

are known as ribonucleic acid and are represented in short-hand terminology as RNA and those containing the second type of sugar are called deoxyribonucleic acid and are given the symbol DNA. Some people use the term *pentose nucleic acid* for RNA and give it the symbol PNA, and this difference in nomenclature sometimes leads to confusion. There are five bases associated with nucleotides of which two (adenine and guanine) are purines and three (uracil, thymine, and cytosine) are pyrimidines.

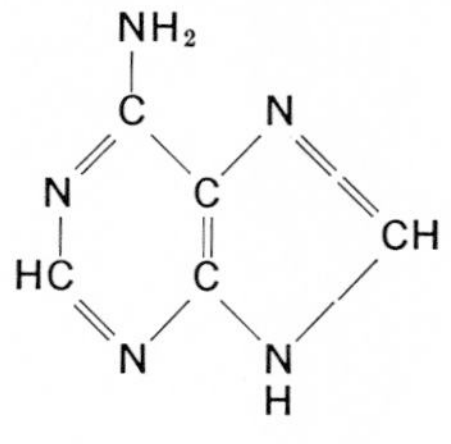

Adenine
(6-Aminopurine)

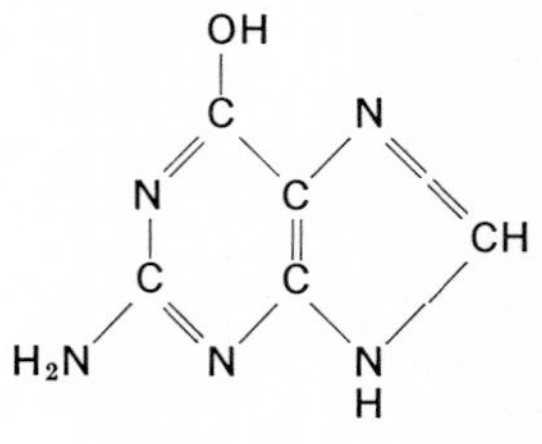

Guanine
(2-Amino-6-oxypurine)

Uracil
(2,4-Dioxypyrimidine)

Thymine
(5-Methyl-2,4-dioxypyrimidine)

Cytosine
(2-Oxy-4-aminopyrimidine)

Originally DNA was obtained from the thymus because of the high concentration of nuclei in such a gland and was called thymonucleic acid, and the RNA was obtained from yeast and was called yeast nucleic acid. It was thus thought that the DNA was characteristic of animals and RNA characteristic of vegetable tissues, however, it has now been found that both DNA and RNA are present in both animal and plant cells but are distributed differently. DNA is concentrated more in the nuclei and this is why DNA was the type of nucleic acid isolated when thymus gland, which has many nuclei, was used as a source.

In the formation of nucleotides, compounds known as nucleosides are first produced. A nucleoside is a molecule that is formed by the combination between a base, which can be either a purine or pyridimine, and one or another of the two sugars, i.e., adenine, guanine, cytosine, uracil, or thymine, associated with a sugar will produce a nucleoside; if they are associated with the ribose sugar, they are one type of compound and with the deoxyribose, they form another type of compound. Adenine, for instance, with ribose forms adenosine and with deoxyribose it forms deoxyadenosine; guanine forms guanosine with ribose and deoxyguanosine with deoxyribose; cytosine becomes cytidine with the sugar and deoxycytidine with the deoxy sugar; uracil with the ribose sugar forms uridine but the

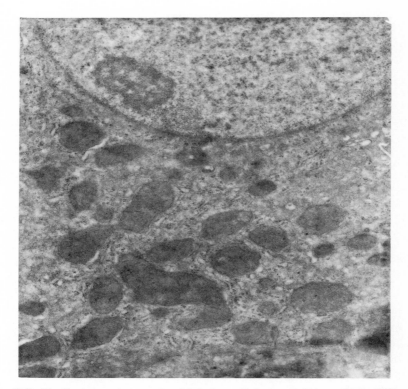

FIG. 97 Electron micrograph of tubular cell from rat kidney. Note RNA granules apparently passing through nuclear membrane to right of nucleolus and in the portion of the nuclear membrane to the right. (Preparation and photograph by R. Q. Cox, Dept. of Anatomy, Emory Univ.)

corresponding compound, with the deoxyribose, is not recorded. Thymine forms thymidine with deoxyribose.

Adenosine (9-β-ribofuranosidoadenine)

The next stage in the formation of a nucleotide is the combination of a phosphate with the nucleoside. For instance, adenosine can have one, two, or three phosphates attached to it and it then becomes adenosine monophosphate, diphosphate, or triphosphate, respectively. Similarly the other compounds mentioned can be mono-, di-, or tri-phosphorylated. Phosphorylation of the five bases with a single phosphate group forms adenylic, guanylic, cytidylic, uradylic, and thymidylic acids and isomers of these are also known. Adenylic acid is commonly spoken of as "muscle" adenylic acid because it was first isolated from muscle. It is one of the structural units of ribonucleic acid and its phosphoric linkage is to the five-carbon atom. In plants, however, the phosphate linkage in the monophosphate appears in many cases to be with the three-carbon atom.

The various nucleotides mentioned in the foregoing become polymerized by the formation of phosphoric acid bonds between the sugars, and polynucleotides are thus formed. The common name for these is "nucleic acids." When the nucleotides are made from deoxyribose sugars, the deoxynucleic (also referred to as desoxynucleic) acids (DNA) are formed and if the ribose sugar is involved ribose nucleic acid (RNA) is obtained.

The structure of these two compounds can be diagrammatically demonstrated,

RNA

Base   Base   Base   Base

  OH   OH   OH   OH

  PO₄   PO₄   PO₄   PO₄

ribose   ribose   ribose   ribose

The bases in each nucleotide vary and may be any one of those first mentioned (i.e., adenine, guanine, cytosine, or uracil). The frequency of each in any particular RNA molecule depends upon the origin of the RNA.

The distribution of DNA and RNA in cells is shown in Fig. 95.

DNA has fundamentally the same structure as RNA but lacks the OH groups shown as attached to the sugars in RNA and uses

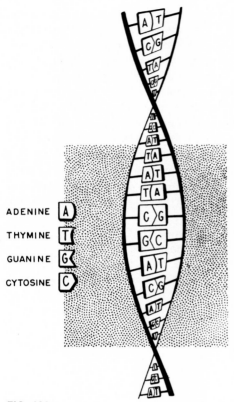

ADENINE A

THYMINE T

GUANINE G

CYTOSINE C

FIG. 98A

FIG. 98 A and B. Section of DNA helix to show base-pair sequences. (Figure 98A from *Med. News,* May 11, 1960, Supplement on Nucleic Acids. Figure 98B from *Rassegna Med.* **37,** 228, 1960.)

the three bases—adenine, guanine, and cytosine but replaces the uracil by thymine.

The formula for DNA looks like this, that is, it is very similar to RNA.

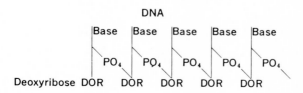

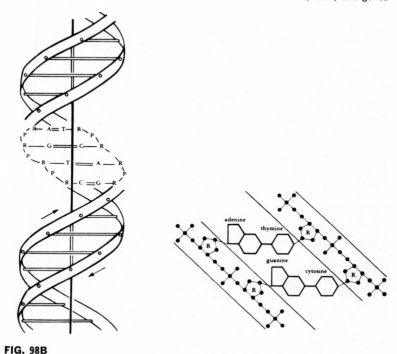

FIG. 98B

Preparations of DNA have been found to have the form of long fibers or chains of nucleotides with a narrow diameter of about 20 Å and are twisted around each other in pairs.

The suggested attachments of the two DNA chains in the spiral is through their individual bases and certain bases couple with specific bases, e.g., adenine and thymine, and cytosine and guanine must be opposite each other (see Figs. 98A and B, 99). The structure therefore is something like this.

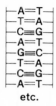

etc.

The order and frequency of bases shown is empirical and would depend on the origin of the DNA.

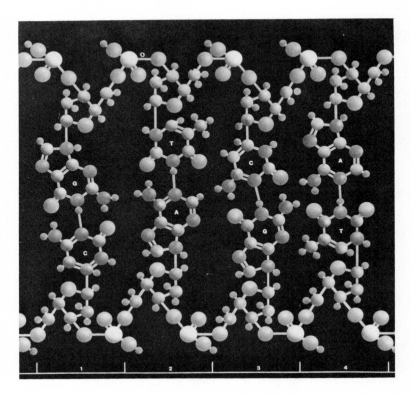

FIG. 99 There are only four arrangements found in the transverse bars of DNA. 1. Guanine–cytosine. 2. Thymine–adenine (TA). 3. Cytosine–guanine (CG). 4. Adenine–thymine (AT). These pairs of bases are repeated thousands of times in DNA in the most widely varied sequences such as T, AT, AT, TA, GC, AT, GC, TA, TA, CG, etc. The "language" of DNA consists of three of these bases which would by analogy correspond to the letters of an alphabet. Still in analogical terms, a word could be formed by the *linear* sequence of the three bases. In the above figure the word formed by the first three bases is GTC, in the lower chain it is CAG. The possible arrangements of the triplets and of their linear sequence in the extremely long DNA chains are so numerous in practice that they explain not only the existence of the living and the extinct species (as well as the differences between individuals of the same species) but also give free rein to nature's whims of millions and millions of other possible species. (From *Rassegna Med.* **41,** No. 3, 1964.)

In the living cell the nucleic acids are inevitably combined with proteins to form nucleoproteins and again, because of the nature of the sugar, one can specify two types of nucleoprotein, deoxyribonucleoprotein and the ribonucleoprotein. The main protein that combines with DNA in these complexes is of a basic nature and is usually either histone or protamine, and there are varying amounts of protein and DNA combined under different circumstances.

It is of interest that viruses and bacteriophages are largely composed of nucleic acids, and in the case of the bacteriophages the DNA can be extruded from the virus particle into the bacterial cell where it can lead to formation of fresh virus; but this will be mentioned again later.

NUCLEIC ACIDS OF THE NUCLEUS

DNA is segregated away from the cytoplasm within the nucleus, although, as we have seen, the isolation of the nucleus from the cytoplasm is by no means complete since there are a number of pores that are present in the nuclear membrane and lead into the cytoplasm. But the significant thing about DNA is that most of it is localized in the nucleus, and it is the DNA that has the major genetic function in the cell (see Fig. 95).

What evidence do we have that DNA is concerned with heredity? The evidence for this has been excellently presented by Brachet in his book "Biochemical Cytology" and will be summarized here.

Seven of the many points in evidence for this are:

1. There is a specificity of DNA's from different species.
2. There is a specific localization of DNA on chromosomes, e.g., in *Drosophila* there is an identity between the DNA bands and the localization of the genes.
3. There is an approximate constancy of amount of DNA per chromosome set.
4. DNA is relatively metabolically stable.
5. Mutagenic agents affect DNA.
6. DNA plays a role in 'phage reproduction.
7. Bacterial transforming agents have been identified with DNA.

With regard to point 1, striking differences were seen in the composition of bases in various DNAs that were isolated from different species of animals. Point 2, the localization of DNA as demon-

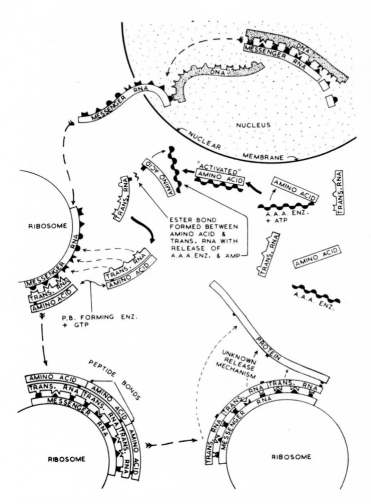

FIG. 100 Diagrammatic illustration of the roles of DNA and RNA in protein synthesis. (From L. T. Threadgold, "Ultrastructure of the Animal Cell," Pergamon Press, New York, 1967.)

strated by the Feulgen reaction and its relationship to the genetic areas, has been demonstrated. Point 3, some constancy does occur, it is not absolute but there is a marked tendency toward such a constancy. Point 4, DNA is probably the most stable of the phosphorus compounds that are found in the cell and is very much

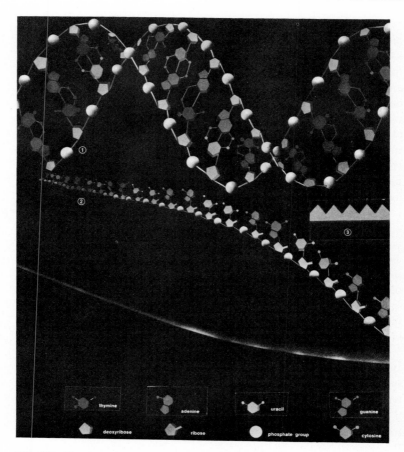

FIG. 101 The genetic code stored in the DNA molecule (1) is transcribed on an RNA molecule (2) known as messenger RNA because it has the task of transferring the genetic message from DNA to the ribosomes where protein synthesis should take place. The synthesis of "messenger RNA" is catalyzed by the enzymatic factor RNA polymerase (3). The arrangement of the bases of messenger "RNA" is complementary to that of DNA bearing in mind that in RNA thymine is replaced by uracil. (From *Rassegna Med.* **41,** No. 3, 1964.)

more stable than RNA. Point 5, all agents that have a mutagenic effect, for example, nitrogen mustard, ultraviolet rays, and x rays, affect DNA. Most of the points quoted so far are circumstantial, but point 6, the role of DNA in 'phage reproduction, seems to provide more direct evidence, as will be explained here.

Some investigators, e.g., Hershey (who shared the 1969 Nobel prize) and Chase, labeled DNA with ^{32}P (radioactive phosphorus) and its protein with ^{35}S (radioactive sulfur). The 'phages used (T2 type) have a head and a tail; the membrane of the head is made of protein and contains DNA (see Fig. 106). When the 'phage is added to the bacteria it infects, the tail of the 'phage becomes attached to the membrane of the bacterium by a submicroscopic sucker. The tail of the 'phage has recently been shown to be cross striated and con-tractile and to contain a solid core. Contraction of the tail shoots the core, like a harpoon, into the bacterium. Attached to the proximal part of the core is one end of a long, single, greatly folded DNA molecule, which occupies the head of the 'phage. The solid core carries this molecule into the substance of the bacterium just as a harpoon carries with it the rope to which it is attached. Practically no protein enters with the DNA, and the DNA then takes over the synthetic process of the bacterial cell and uses its metabolism to produce some virus protein and a good deal of its own DNA instead of bacterial protein. Thus the DNA of the 'phage is able to produce in the bacterium DNA that is genetically identical with that of the infecting bacteriophage and genetically quite different from the DNA of the bacterium (see Figs. 105–108).

Point 7, the significance of DNA for bacterial transformations, is explained as follows. There are two types of pneumococci, one type known as the encapsulated or smooth strain and a second type, which is not capsulated and is known as the rough strain. If DNA from the first of these bacteria is brought into contact with the rough strain of *Pneumococcus*, the latter will be converted into the smooth strain and this property will be transmitted to the de-scendants of the transformed strains and these descendants will produce DNA with the same properties. Thus the DNA is found to be directly concerned with the synthesis of a particular type of capsule. Bacterial transformations have also been obtained with a number of other bacteria using DNA.

This work has suggested that at least some and possibly all the genes found in bacteria were probably completely or largely composed of DNA.

It is of interest that bacterial viruses (bacteriophages) are all DNA viruses, that plant viruses are all RNA viruses, and that some human viruses contain RNA and some contain DNA (see Figs. 105–108).

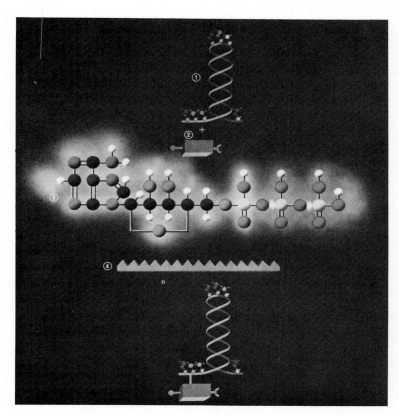

FIG. 102 In addition to "messenger RNA" there is "transfer RNA" (1). Its task is to convey amino acids (2) in the correct sequence to the "messenger RNA" arranged on the ribosomes. There appears to be a separate type of "transfer RNA" for each of the 20 common amino acids. The groups of bases in the RNA help to distinguish the exact site where the transported amino acid should be placed. The hooking process of the "transfer RNA" with a specific amino acid requires the presence of a specific enzyme (4). The energy necessary for the process is supplied by ATP (3) (The large molecular structure at the center of the figure.) The energy available from ATP is stored in the last two phosphate groups and is depicted as a wavy line between the groups. (From *Rassegna Med.* **41**, No. 3, 1964.)

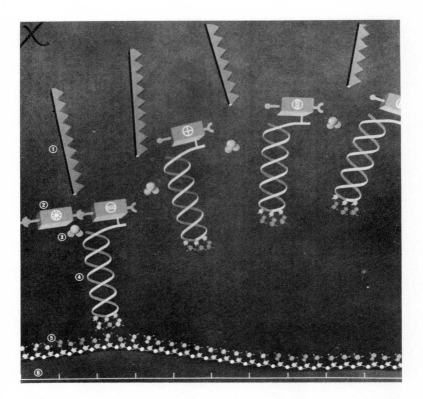

FIG. 103 Having left the nucleus, the "messenger RNA" (5) deposits on the ribosome (6); the sequence of the bases specifies the order which the amino acids must have in the protein to be synthesized. Meanwhile "transfer RNA" (4) after hooking the various amino acids (2) carries them to the proper sites. Once the amino acids have reached their site, they must be linked together to form a protein. The union between one amino acid and another is achieved with the aid of enzymatic factors; (1) the formation of the peptide bond gives rise to a molecule of water (3). (From *Rassegna Med.* **41,** No. 3, 1964.)

It has previously been mentioned that the model for DNA was a two-stranded molecule arranged in a helical form with the pairs of bases held together by hydrogen bonds, for example, adenine would combine with thymine and guanine with cytosine (see Figs. 96 and 98A and B). This is a sort of zipper arrangement and if these pairs of molecules could be unzipped by breaking the hydrogen bonds, then each of the members of the chain could serve as a template for the manufacture of a second chain. This is the type of molecular arrangement that would permit a gene to reproduce itself, and there is very strong evidence that DNA molecules could, in fact, be equated with genes. Although DNA fulfills these requirements, we cannot be sure, of course, that when it is incorporated into the cell it simply exists as DNA. Most likely it combines with a protein that probably exerts an influence on the DNA in its function as a gene.

The synthesis of protein in a cell is attributed largely to RNA, but the code for the sequence of the arrangement of amino acids in the protein being synthesized is communicated to the RNA by the DNA (see Figs. 100–103).

Current theories are that one of the strands of DNA in the double helix forms RNA and that the order of its nucleotides is determined by those in the DNA strand. The RNA so formed is "messenger" RNA, which passes out of the nucleus to the cytoplasm, and according to some authors it is combined with protein either before or after leaving the nucleus to form a ribosome. Protein is formed on the RNA of the ribosome, the sequence of amino acids being in accordance with the sequence of nucleotides in the RNA. The amino acids are brought to the ribosome attached to transfer RNA. The interrelationships of these RNA's with synthesis of proteins are shown in Figs. 100–103.

The RNA copy of the DNA strand is believed to be produced at the same time that the double helix unwinds so that its synthesis occurs on a single DNA strand. A new theory of how this acts has been suggested by Dr. Raymond Smith of the University of Indiana. He suggests, according to *Scientific Research* (October 13, 1969, p. 27), that the transcription of RNA from the DNA consists of four steps.

1. The DNA double helix is attacked by a complex package of enzymes, known as endoenzyme. This complex consists of (a) an exonuclease to detach the deoxyribonucleotides, (b) an RNA polymerase molecule to catalyze the formation of messenger RNA, and (c) an enzyme that links up again the two DNA strands of the double helix.

2. The endoenzyme package does not unwind the DNA strands of the double helix; it simply detaches a single deoxyribonucleotide from one DNA strand.

3. The detached deoxyribonucleotide is used as a template for the incorporation of the appropriate ribonucleotide into messenger RNA.

4. The deoxyribonucleotide is replaced in its original position in the DNA strand.

Genetic information tends to exert its effect despite the subjection of the organism concerned to variations in temperature, salinity of the fluids with which it comes in contact, different types of food, the amount of light it receives, exposure to toxic substances, and so on. However, radiation is one environmental factor which can exert a profound effect on the genetic material, and it affects particularly the DNA.

It is of interest that in the passing on of genetic information to the organism by the DNA there is no feedback of information, so far as we know, to inform the DNA of what has been done or what needs to be done but, despite this, practically all significant cell activity is affected by the gene.

If proteins are hydrolyzed they break down to amino acids and, under natural circumstances, a protein is made up of amino acids joined together in the form of elongated chains known as polypeptide chains. In a protein of molecular weight of about 25,000 there will be something like 230 residues joined end to end to form the single polypeptide chain of which the molecule is made. The amino acids are invariably joined together in this chain by the same method. The bonds are covalent bonds, and the amino acids are joined together by the formation of a peptide link that results in the elimination of a molecule of water; nearly all covalent links in a protein are formed by this method, however, occasionally there are links between two sulfur molecules.

The protein so formed therefore is an elongated linear molecule, and there is no evidence that it is branched. There are only about 20 different kinds of amino acids in proteins, but these same 20 occur, generally speaking, in practically all proteins—it does not matter whether the proteins come from animals or plants or microorganisms. Not all proteins contain all these amino acids but all proteins do contain some of them. Another point is that all the amino acids present in proteins have the L-configuration.

There are some specific amino acids that occur only in specific proteins, and probably the best example of these is hydroxyproline, the amino acid that is characteristically found in collagen. The 20 amino acids found universally in proteins are as follows: glycine, alanine, valine, leucine, isoleucine, proline, phenylalanine, tyrosine, serine, threonine, asparagine, glutamine, aspartic acid, glutamic acid, arginine, lysine, histidine, tryptophan, cysteine, methionine. Other amino acids that may be present are: hydroxyproline, hydroxylysine, phosphoserine, diaminopimelic acid, thyroxine, and cystine. F. H. C. Crick points out that not only is the composition of a particular protein fixed but it appears that the order of the amino acids in the polypeptide chain is determined very exactly; for example, every molecule of hemoglobin in human blood has precisely the same sequence of amino acids as every other molecule of hemoglobin. A further point is that the polypeptide chains of the proteins are not extended but are folded upon each other, and this folding is said to be maintained by physical bonds that are weak in nature and possibly also by disulfide linkages and other linkages. The degree of folding is thought to be similar for each particular protein, and presumably the denaturation of proteins results in the destruction of this folding. If any one of the 20 essential amino acids is supplied to a cell, it can be incorporated into proteins but if, for example, any one of the 20 is not available to an organism, protein synthesis cannot occur. Not only is the synthesis of that part of the protein molecule which does contain the specific amino acid prevented but also that part of the molecule that does not contain the amino acid is also not synthesized. Crick points out that, with this type of mechanism and with the order of the amino acids being very specific, one would expect that some variation or mistakes in order would occur occasionally, but the information we have at the moment is that such mistakes are very infrequent. According to J. T. Bonner of Princeton University an average sized cell contains about 2×10^{14} molecules of which about 15% are proteins. About 50,000 different proteins have already been identified. The number of proteins that could be synthesized in nature is very large indeed, and by various combinations and by combinations in different ways of the 20 known amino acids, it would be possible to have 10^{1278} protein polymers, all with different biological properties.

Although all the hemoglobin molecules of the human being are the same, if this hemoglobin is compared with that of the horse or some other animal, it will be found that there is a general similarity of pattern between the two molecules and that the amino

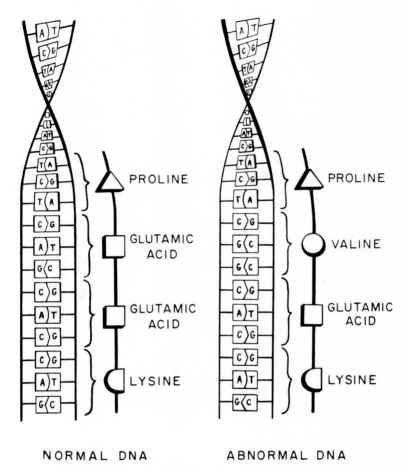

NORMAL DNA   ABNORMAL DNA

FIG. 104 Synthesis of normal and sickle-cell hemoglobin. (From *Med. News*, May 11, 1960, Supplement on Nucleic Acids.)

acid composition will be pretty much the same for both of them. These two hemoglobin molecules may differ a little in their elec- trophoretic properties—their crystalline form may differ and the ends on their polypeptide chains may be different, but it is very likely that the sequence of amino acids in the polypeptide chains will be fundamentally the same except for one or two slight altera- tions in the sequence. It is of interest that this produces what might be described as a family likeness between proteins, and Crick suggests that it may be that "these sequences are the most delicate expression possible of the phenotype of an organism and that vast amounts of evolutionary information may be hidden away within them."

It was found that the nature of the protein in the hemoglobin in human sickle-celled anemia is different from that of normal humans. This disease has been described by Pauling as a "molecular disease." Ingram (1956, 1957), has shown that this difference results from the fact that valine replaces glutamic acid in the DNA chain (see Fig. 104) and this, according to Ingram, is the only change present in the molecule. Crick says of this, "It may surprise the reader that the alteration of one amino acid out of a total of about

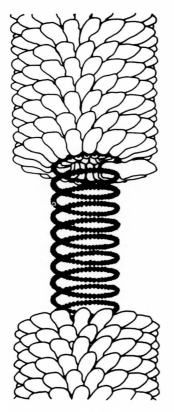

FIG. 105 Helical arrangement of repeating coat protein in a tubular virus. A portion of coat protein has been cut away to expose inner helix of DNA. (From Caspar and King, *Cold Spring Harbor Symp. Quantitative Biol.* **27**, 1967, *in* "Molecular Insight into the Living Process" (D. E. Green and R. F. Goldberger, eds.), Academic Press, New York, 1967.)

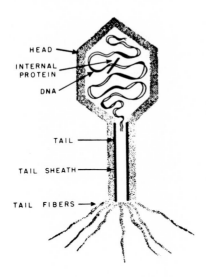

HEAD

INTERNAL PROTEIN

DNA

TAIL

TAIL SHEATH

TAIL FIBERS

FIG. 106 Infection of bacteria with T_2 virus particles. Diagram of single virus. Note DNA in the head. (From *Med. News,* May 11, 1960, Supplement on Nucleic Acids.)

300 can produce a molecule which (when homozygous) is usually lethal before adult life.''

When the process of synthesis of proteins is studied, one particularly important problem has to be considered and that is how to explain the mechanism that controls the order of the amino acids in a protein. The condensation of amino acids into polypeptide chains is fairly simple to explain chemically but their order is not. This order must be very critically and stringently controlled since, as we have seen, the slightest variation in sequence of amino acids in the hemoglobin molecule can produce a lethal disease. Thus the critical point in protein synthesis is the joining up of the amino acids in a predetermined order. It is now known that DNA controls the sequence of amino acids in a protein through messenger RNA.

That DNA can affect protein synthesis is demonstrated by the fact that when DNA is squirted by the T2 bacteriophages previously described into bacterial cells without any of the protein going in with it, it appears to be able to control the synthesis of protein inside the bacterium.

The presence of pores in the nuclear membrane would seem to indicate that it would be freely permeable to quite large molecules; however, the nuclear membrane shows osmotic properties and the pores must therefore not be simple holes but must have some differential control over the compounds that pass through them. Some

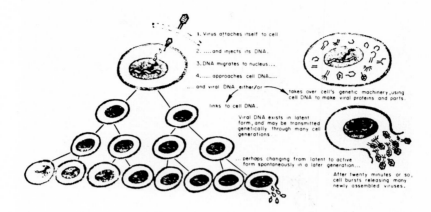

FIG. 107 Results of T₂ virus invasion. Note how DNA of virus diverts metab-olism of bacterium into reproduction of virus DNA and protein. (From *Med. News*, May 11, 1960, Supplement on Nucleic Acids.)

workers have talked about a thin membrane covering the pore, others have found a sort of electron dense ground plasm filling it that appears to contain protein, and of course, as one might expect, a number of authors have discovered RNA particles in them. More recent studies indicate that the actual functional pore is much smaller than the pore demonstrated by the EM and it seems that even these tiny openings may be temporary in nature. Feldherr [*Exptl. Cell Res.* **38,** 670 (1965)] found in the ameba that gold particles smaller than 85 Å would penetrate the nuclear mem-brane in a period of 3 min, whereas particles bigger than that had difficulty in getting through at all. The pores in the nuclear envelope of the ameba measure about 640 Å in diameter; thus there cannot be a hole of this size available for passage of molecules. The actual size of the available channel has been estimated to be about one quarter of the pore diameter. In the nucleus of oöcytes of the frog, there is evidence that there is some electrostatic control of the passage of substances through the pore, negatively charged particles, for example, tending to stick to the periphery of the pore. Du Praw [*Proc. Natl. Acad. Sci. U.S.* **53,** 161 (1965)] has produced evidence that the size of the pore may vary and have a sphincterlike action. This is a suggestion that would explain many of the anomalies and perplexities of nuclear membrane penetration.

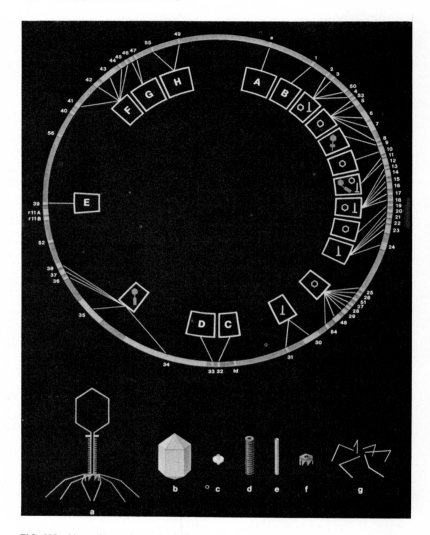

FIG. 108 Hereditary characteristics are transmitted by nucleic acids, even in viruses. The ring-shaped DNA of a bacteriophage (a) is shown schematically. Under the electron microscope the virus can be seen to be composed of a head (b), a collar (c), a sheath (d), and a central axis (e). Other structures have also been detected, e.g., a plaque (f) and basal filaments (g). The information necessary to construct the protein capsule of a bacteriophage and its auxillary structure is contained in a single chromosome enclosed in the head. The DNA molecule in the chromosome is round and geneticists have succeeded in identifying numerous gene loci on it. Some carry the structural

One of the subjects that has intrigued scientists has been the precise mechanism whereby DNA influences the sequence of amino acids in protein. There is now substantial evidence that this sequence is decided by a code built into the DNA, a kind of language. Dr. G. W. Beadle, who won a Nobel prize for his discovery of the regulating effect of genes on chemical events, points out that this kind of language is at least 3 billion years old, it is written as Beadle points out in a "submicroscopic molecular code." In an article in *Rassegna Medica,* Beadle says of the code "it has only four letters (adenine, thymine, cytosine and guanine). Each of these is one two hundred millionths of an inch across. Each is made of but five kind of atoms—carbon, hydrogen, oxygen, nitrogen, and phosphorus, and there are only about thirty two atoms in each letter. All the words in this language are three-letter words and only sixty four of them are possible. This beats basic English by a very wide margin. Most remarkable of all, this language is the only language known to have a built-in capacity for copying itself. There is no need for secretaries, typewriters, or printing presses. It is a language in which the specifications for all living organisms, from virus to man that have ever existed on earth have been written.

"The number of letters in the specifications for these organisms varies from five thousand letters for the specifications of the simplest known virus to five billion for the specification of man. That's over a million fold difference. What is the name of this language? I call it DNAese, because it is contained in deoxyribose nucleic acid molecules, which we call DNA."

It has been shown earlier that a single strand of DNA has a series of bases associated with it; these are adenine, thymine, cytosine, and guanine and they may be arranged in any order along the length of the DNA molecules. There is evidence now that a series of any three bases can form a "word" of the DNA language.

---

characteristics of the protein capsule, others contain information about the physiology of the bacteriophage lysozyme production, the synthesis of new viral DNA. Geneticists have also succeeded in altering certain loci and have thus obtained specific mutations involving one or more of the elements making up the bacteriophage. The altered elements are light gray in color on the chromosome. (A) No lysosome production. (B) No DNA production. (C) No DNA production. (D) No bacteriophage maturation. (E) Delayed DNA production. (F) DNA production halted. (H) No bacteriophage maturation. (From *Rassegna Med.* **45,** No. 1, 1968.)

Thus if we use the initials of the individual bases we can have the following examples of words ATC, TAC, CTA, GGA, CGA, AGC, and so on. The possible arrangements of these three-letter words and their sequence along the DNA molecule is enormous. Beadle points out that in the nucleus of the egg from which we are all developed the molecules that constitute the various letters if strung along side each other would form a DNA chain 5 ft long and this would contain 5 billion "letters," or 1 billion seven hundred thousand three-letter words.

Watson and Crick (the former is the author of "The Double Helix") received a Nobel prize for showing that the DNA molecule is really a double chain of molecules (letters) arranged in a helical pattern (see above). The chains are linked together by means of hydrogen bonds between the bases of the nucleotides. The phosphates and sugars are actually located outside the chain while the bases are located inside it. As mentioned earlier it has been shown that only adenine and thymine and cytosine and guanine will link each other; adenine will not link up with cytosine or guanine and neither will thymine. If therefore the order of bases is known in one chain, the order of bases in the complementary chain will also be known. If, for example, order of bases in one strand of the DNA is ATCAGTCA, the order of the bases in the complementary strand is TAGTCAGT (see diagram on page 207). The DNA chain therefore replicates itself by separating into two strands and each of the strands then regenerates a complementary strand.

Arthur Kornberg, Nobelist, of Stanford University has demonstrated that in the test tube a single strand of DNA will replicate itself if the appropriate nucleotides are also present. DNA, through the intermediary RNA, dictates the synthesis of protein. Protein molecules, as has already been indicated, are large molecules made up of smaller molecules known as amino acids, and, as previously indicated, there are 20 different kinds of these acids. Their different combinations build up the tens of thousands of proteins that exist in nature, many of which have been identified. Apart from water the major constituents of soft tissues of the body are made of protein— even the enzymes of the cells are made of the same material.

In the synthesis of protein, a DNA strand in the nucleus replicates an RNA strand that is complementary to it. It is important to note, however, that the base thymine in DNA is replaced by the base uracil in RNA. The RNA (messenger RNA) then leaves the nucleus and becomes incorporated into a ribosome where it serves as a template where amino acids are strung together to form a protein. It may rightly be asked how the RNA serves to act as a template for the serial orientation of amino acids. If we take a series of three nucleotides on a strand of DNA with bases ATC in that order, then the RNA that replicates from that strand will have nucleotides with bases TAG in *that* order. Now it happens that each three-letter word (composed of three nucleotides) will be equivalent to a specific amino acid. A series of three-letter words in series along a strand of RNA will lead to the fixing of three amino acids in order and so on. The amino acids are brought to the ribosome attached to a section of RNA known as "transfer RNA." There are 20 different sorts of transfer RNA and each one carries a three-letter nucleotide "word" that "fixes" the individual amino acid that it carries. This three-letter word is the complement of a word on the messenger RNA, so the two complementary words will lock together at a particular position on the messenger RNA. The various amino acids are therefore brought side by side in the right order. The amino acids then link up to form a protein.

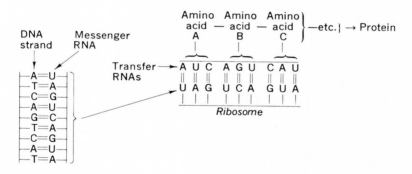

It will be noted that the transfer RNA has the same bases in the same sequence as in the original DNA strand. Thus the DNA presides over the protein that forms the major part of the soft material of the body and forms the enzymes that control the metabolism of the body. Sometimes a defect in the base sequence results in the failure to produce a key enzyme, and so-called "genetic disease" is produced. In a sense all primates suffer from a genetic disease. This is the lack of the ability to synthesize vitamin C, which has therefore to be included in the diet. With one or two exceptions (e.g., guinea pig) all other mammals can synthesize this vitamin. Primates can bring the synthesis of vitamin C to the last stage, but in the process of evolution the gene controlling the formation of the enzyme (gulonolactone oxidase) responsible for the final stage of the synthesis dropped out, and primates then had to get their vitamin C from their food if they were to survive. Similarly, the disease phenylketonuria, which produces feeble-minded children, is caused by a change in the gene that produces the enzyme that oxidizes the amino acid phenylalanine to the amino acid tyrosine. Normal proteins contain phenylalanine and the body reconverts some of this back into protein. In the absence of the enzyme, this excess phenylalanine is not converted and accumulates in the body and exerts a toxic effect on the central nervous system.

The difference in the DNA coding for the proteins of hemoglobin in normal blood and sickle-celled anemia is shown in Fig. 104. Change in the coding for one amino acid—glutamic acid—so that valine takes its place, results in change in the hemoglobin that makes the difference between normal blood and a disease. Some hundreds of genetic diseases are known in man.

Sometimes mistakes occur in the replication of the DNA molecules and if this happens in a germ cell, then a new character will develop—a "mutation." If it is a favorable one, the animal will survive; if it is unfavorable, the animal will be eliminated. In modern society unfavorable mutations such as that which produced phenylketonuria can be treated, and the individuals survive to pass this gene on to their offspring.

We see that three-letter words represent the code for certain amino acids, e.g., AGA (adenine, guanosine, adenine) is the code for the amino acid—"Proline" GTC (guanosine, thymine, cytosine) for glutamic acid—GCC for valine, and GTC for lysine. How do we know that these particular three-letter words are the keys to specific amino acids? One of the leading experts in this field is Nobel prize winner Severo Ochoa who worked especially with RNA. We have seen that RNA differs from DNA mainly in the fact that the associated sugar is ribose instead of deoxyribose. It also contains uracil, whereas DNA contains thymine (5-methyl uracil).

Dr. Ochoa made RNA artificially in the test tube using an enzyme that he had isolated called polynucleotide phosphorylase. If, for example, he used nucleotides all with the same base, he was able to produce RNA that had only those bases in it. If he put in two types of nucleotides, he got RNA with the bases of both types of nucleotides in it. Later on Drs. Nirenberg and Matthai of the National Institutes of Health who received a Nobel prize in 1968 put mixtures of amino acids into a test tube and then added artificial RNA made up entirely of uracil and produced a "protein" made up entirely of phenylalanine (more correctly it was a polyphenylalanine chain). When they put in an RNA made up only of nucleotides containing uracil and cytosine in 5 : 1 ratio, the chain formed a polymer (or "protein") made up of four amino acids, namely phenylalanine, leucine, serine, and proline. Now uracil and cytosine can form the following three-letter "words" or triplets—UUU, UUC, UCU, CUU, UCC, CUC, CCU, and CCC. The studies with RNA containing only uracil showed that the UUU would incorporate phenylalanine, similar studies with RNA containing only cytosine showed that the CCC triplet would incorporate proline. The other triplets could be worked out from the percentage of incorporation of the various amino acids. Further studies have shown that a number of amino acids are coded by more than one three-letter word (triplet). This is known as degeneration of the code. This is why proline is shown coded by AGA on one strand but TCT on the first strand, whereas above we showed it was coded if uracil was the base replacing thymine in RNA to UCU. No amino acid needed four bases for its incorporation, and this is why we say that the language of DNA consists of three-letter "words." Subsequent studies by Nirenberg and Leder showed that AAA binds lysine, UGU binds cysteine, UUG binds leucine ACU binds threonine, GCU binds alanine, GGU binds glycine, AGA binds arginine.

Until recently it was assumed that all the DNA that carried genetic information was located in the nucleus. However, certainly cytoplasmic components seem to lead an existence in which they grow and even reproduce without apparent nuclear direction. Both mitochondria and chloroplasts came into this category, and there is a school of thought that we have mentioned in the chapter on mitochondria that believes that both these structures are in fact independent organisms that became associated with cells. In 1964 DNA was isolated from mitochondria, but as mentioned earlier it was an unusual form of DNA being arranged in a circular fashion with no free ends and very different from the Watson-Crick elongated helical structure of DNA. This mitochondrial DNA may also be twisted around the center of the circle like a many-twisted figure 8.

There is now strong evidence that the synthesis of mitochondria in the cell is dependent partly upon the DNA of the nucleus operating through messenger RNA and the cytoplasmic ribosomes. The DNA of the mitochondria appears to be located in the matrix. The inner membranes of the mitochondria contain enzymes for the synthesis of DNA and RNA and also a mechanism for the synthesis of protein that appears to be different from the extramitochondrial systems.

In 1967, Drs. B. and G. Attardi [*Proc. Natl. Acad. Sci. U.S.* **55** (1967)] announced that they had actually obtained evidence that a messenger RNA was produced in mitochondria. Using radioactive isotopes they found that some of the isotope labeled RNA turned up in the cytoplasm faster than RNA that came from the nucleus. This new RNA was a molecule of different size from that produced by the nucleus and different in chemical composition. Furthermore, this new RNA was found to be associated with a section of the rough endoplasmic reticulum that is regarded by some workers as being concerned with the synthesis of the membranes of the cytoplasmic organelles. The new RNA was also found to have the circular configuration similar to the mitochondrial DNA, and to have a base sequence complementary to that of the mitochondrial DNA. This work was carried out on human HeLa tumor cells thus indicating that cytoplasmic inheritance as represented by the mitochondrial DNA exists in organisms as high on the evolutionary scale as humans. It is of interest that in cell division the mitochondria

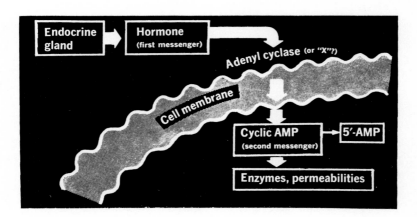

FIG. 109 Mechanism of effect of hormones on cell enzymes (see text). (From *Sci. Res.*, October 16, 1968.)

are divided more or less equally between the two daughter cells and that in fertilization the middle piece of the sperm, which is carried into the egg, is formed at least in part from mitochondria.

During 1968 an interesting series of studies were reported that indicated how the hormones of the endocrine glands might influence the metabolic processes of the cells. The key substance in this mechanism appears to be cyclic adenosine 3'5' monophosphate (see Fig. 109). This is in fact a cyclic nucleotide instead of being a straight nucleotide (one of the constitutents of RNA and DNA), and it was discovered by Drs. Butcher and Robinson of Vanderbilt University in Tennessee. There is evidence now that various hormones stimulate the formation of cyclic AMP in the cells of its targets and that cyclic AMP affects certain enzyme systems in the cells (see Fig. 109). Two enzymes in a cell control the amount of cyclic AMP present; one is adenyl cyclase that catalyzes the formation of cyclic AMP and the other a phosphodiesterase that, by converting cyclic AMP to adenosine 5' monophosphate, inactivates it. The various hormones stimulate one or the other of these enzymes in the appropriate target organs. At least adenyl cyclase appears to be located in the cell membrane so it can obviously be affected easily by hormones. Among hormones that appear to work by stimulating adenyl cyclase to produce cyclic AMP are the adrenocorticotrophic hormone (ACTH), the luteinizing hormone, angiotensin, gastrin, histamine, and serotonin. Some hormones act by depressing the level of cyclic AMP in the cytoplasm. These include insulin, melatonin, and the prostaglandins. Some hormones such as the catecholamines (from the adrenal medulla) can increase the cyclic AMP in some tissues and decrease it in others. Also the sensitivity of different tissues to cyclic AMP varies considerably; for example, the activity of the enzyme phosphorylase in the liver is greatly affected by it, but the phosphorylase of the brain is not affected at all. In the cell itself cyclic AMP causes the following changes: an increase in the enzymes phosphorylase and phosphofructokinase, release of insulin (pancreas), release of protein from ribosomes, increase in kidney permeability, and increase in HCl secretion from the gastric mucosa. On the other hand the following are decreased— glycogen synthetase, conversion of acetate to liver cholesterol, tone of smooth muscle, incorporation of amino acids into protein in the liver. It can be seen, therefore, that at least in the liver it has a direct connection with protein synthesis causing a release of already formed protein from the ribosomes but slowing down the formation of new protein.

Up to the present we have been considering mainly the synthesis of protein in the cytoplasm; we know that both DNA and RNA are present in the nucleus and that DNA controls protein synthesis. Perhaps we should divert for a minute here to consider a case where the control of heredity is in the hands of RNA, there being no DNA present. This is the work that has been carried out recently on tobacco mosaic virus by two groups, that of Gierer and Schramm (1956) and the group of Fraenkel Conrad. There are a number of strains of tobacco mosiac virus, and some of these strains have a protein that contains histidine. Gierer and Schramm showed that if you take the RNA part of the virus and free it completely from protein, it can still be infective although it is not as infective as when combined with protein. Fraenkel Conrad was able to recombine the protein and RNA of the virus and obtain again an active virus in which the infectivity was high but not quite normal. It is of interest that by swapping the RNAs and proteins from different strains it is possible to obtain different combinations. If, for instance, the virus, which is made by taking the RNA of one strain and the protein from another, is used to infect a tobacco plant, the virus that is reproduced in the plant is similar not to the virus from which the protein was taken but to the one from which the RNA was taken. So that, for example, if in the particular strain of virus from which the RNA was taken there was no histidine in the protein, then the virus removed from the tobacco plant after infection would also have no histidine; in other words, the RNA of the virus determines the composition of the protein of the virus. Furthermore, the protein that was in the original virus was reproduced in the plant, and this protein was that of the virus from which the RNA originally came.

SYNTHESIS OF PROTEIN BY NUCLEOLI

It has already been mentioned that the nucleolus contains a good deal of RNA but it also contains some DNA, and Caspersson claims it is rich in diamino acids. Nucleoli are known to contain protein, e.g., histones have been found in them and there is evidence that at least five different proteins occur there. The presence of protein masked phospholipids has also been claimed, and alkaline glycerophosphatase and acid phosphatases have been demonstrated in this organelle by a number of authors, and also a great variety of other phosphatases by the present author. Other enzymes are present in the nucleolus, e.g., dipeptidase, cytochrome c reductase, nucleoside phosphorylase, and the DPN synthesizing enzyme; the two latter are present in a higher concentration in the nucleoli than in the cytoplasm. Many SS— and SH— groups are also present, and, in general, the nucleolus has from about 40 to 85% of dry matter, which is, relatively speaking, a very considerable amount. Although the nucleolus is an area in the nucleus where there is a large concentration of special substances, the electron microscope has to date given no evidence that it is surrounded by a limiting membrane. Originally Caspersson suggested that the nucleolus was the center of protein synthesis, and part of this suggestion is due to the fact that the nucleolus is large and well developed in protein synthesizing cells and that in such cells it is particularly rich in RNA.

The conception that the nucleolus plays a part in protein metabolism has been made likely by the work of Ficq. She noted by track autoradiography that glycine and phenylalanine containing radioactive carbon are incorporated more rapidly into the nucleoli than into the cytoplasm. In the case of oöcytes that are growing rapidly and thus synthesizing much protein, the rate is extremely fast. Similarly nucleic-acid precursors are concentrated from 100 to 1000 times faster in nucleoli of such eggs than in the cytoplasm.

Further mention should be made of the localization of alkaline phosphatase in nucleoli. This was first recorded by the present author in 1943 as a result of histochemical studies and was subsequently confirmed by a number of workers—since then the present author has described the remarkable dephosphorylating activity of the nucleoli for a wide variety of phosphate substrates. This is shown particularly by the Purkinje cells of the cerebellum and by other cells which produce appreciable amounts of protein. The nucleoli of these cells readily dephosphorylate glycerophosphate, numerous sugar phosphates, riboflavin-5-phosphate, pyridoxal phosphate, ethanolamine phosphate, various steroid phosphates, and a wide range of phosphate esters that are concerned with intermediary metabolism and particularly the high-energy phosphates. In cells that are actively producing protein, phosphatase is not only present in the large nucleoli but in the cytoplasm of the cell as well. Bradfield, for example, showed that an active phosphatase is present in the cells of the silk-spinning glands of insects that synthesize and secrete protein at an extraordinarily fast rate. Phosphatase is also present in the cells and fibers during proliferation of connective tissue after wounding. There is an increase in the enzyme in the fiber-producing cells at the time of fiber production. In periosteal fibrosarcomas where no bone is produced and also in polyostotic fibrous dysplasia, alkaline phosphatase is seen in both nucleoli and cytoplasm of the fiber-producing cells. Siffert has pointed out that the frequent association of alkaline phosphatase with the matrix of both fibers and cartilage would indicate an association with matrix production. It is of interest that phosphatases are present in the uterus of the hen and the mantle edge (which secretes the shell) of mollusks. Both the uterus and the mantle edge produce considerable amounts of protein. The eggshell of the hen has no calcium phosphate in it, and the shell of the mollusk has only very small amounts of phosphate, so that it seems in these cases the enzyme is more concerned with matrix production, in other words with protein synthesis than with mineralization; the enzyme is also present particularly in regions where histogenesis is occurring. Jeener has shown that alkaline phosphatases are associated with cell proliferation in organs stimulated by sex hormones. Some phosphatases are also present in other parts of the nucleus but their significance will not be discussed here. See Fig. 110 for illustration of extracellular protein produced by fibroblasts.

These facts together with the association of phosphatases with the nucleolus strongly support the concept of the importance of the latter in protein synthesis.

Although the increase of phosphatase in the nucleolus of protein-secreting cells may simply indicate increased RNA production, which may be passing out into the cytoplasm as messenger RNA, it is of interest that there is a localized region of the nucleus (the nucleolus) without an obvious membrane in which, presumably, the protein metabolism of the nucleus is centered. Although the nucleolus is itself separated from the cytoplasm by the surrounding nuclear elements, as mentioned earlier it is mobile and able to make direct physical contact with the nuclear membrane on occasions. It is possible, that on these occasions the nucleolus may discharge its protein or its RNA, or both, through the nuclear membrane. Whatever the role of the nucleolus in protein synthesis, we must accept the conception that a substantial amount of the protein synthesis of the cell takes place in the cytoplasm.

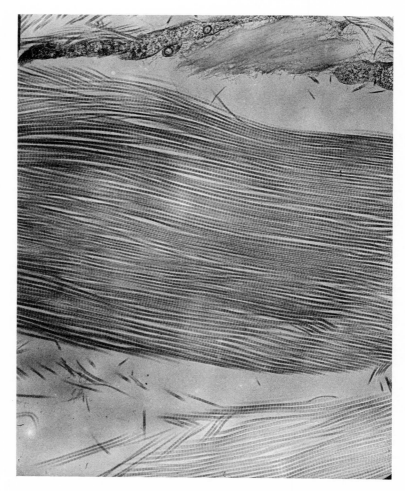

FIG. 110 Electron micrograph of **extra cellular** protein. Collagen fibrils produced by protein synthetic mechanisms of fibroblasts.

Cytoplasmic localization of RNA, as has been demonstrated by many workers, provides evidence for this. The work of Barrows and Chow on the intracellular distribution of vitamin B_{12} adds support to it. Vitamin B_{12} is associated with protein synthesis possibly through the formation of RNA, but RNA *and* DNA are decreased in vitamin B_{12} deficiency, and the incorporation of ^{32}P into nucleic acids is affected in this condition. The distribution of vitamin B_{12} in homo-

genates of the liver is 40% in the microsomes, 13% in the mitochondria, and 11% in the nuclei, with 22% in the supernatant.
Although only 11% of vitamin B_{12} is present in the nuclei, if this were
all localized in the nucleolus it would represent an appreciable
concentration in this region.

It is of interest in view of what has just been said about the
nucleolus that protein synthesis takes place in the cytoplasm even
in the absence of the nucleus. It appears that the RNA particles
(ribosomes) that are associated with the endoplasmic reticulum are
the center of such activity and it is noteworthy that in the synthesis
of protein the amino acids have been said to "flow" through these
particles. The first experiments on this subject were those by
Zamecnik *et al.* (1956). They showed that if rats were given fairly
large quantities of radioactive amino acids, later, after the animal
was killed and liver homogenates prepared, the microsomal particles
(which include the ribosomes) contained a constant amount of this
particular amino acid indicating that they were fixed by this part
of the cell. However, in the second experiment a very small amount
of radioactive amino acid was administered, and, in this instance
within a short time, the radioactivity present in the microsomes
rose steeply and then fell away quite quickly. This is obviously
what one would expect if the protein in these particles was turning
over very rapidly. This means that the amino acids are adsorbed
onto the ribosomes, converted into proteins, and then passed out
of the microsomes again as protein.

As indicated earlier, the amino acids are attached to transfer
RNA, which carries them to the ribosome where they are condensed
into proteins. An amino acid-activating enzyme has been discovered
and found to be widely distributed and is believed to be essential
to all cells that undertake the synthesis of protein. If amino acids
are labeled with radioactive atoms, then the RNA becomes labeled
and can be extracted from the solution, purified, and then added
to the microsomal fraction of a homogenate. The labeled amino
acid is then transferred from the RNA to the protein of the ribosome.
This is an extremely interesting step in the problem of protein synthesis. It is also of interest that RNA can be synthesized without
protein necessarily being synthesized at the same time. Normally
one finds that in a cell that is undergoing active RNA synthesis,
active protein synthesis is taking place simultaneously. However,
the synthesis of protein in such a system can be brought to a

full stop by the use of chloramphenicols, this is particularly well shown in systems obtained from bacteria. If the protein synthesis is stopped in this way by chloramphenicol, synthesis of RNA continues unaffected. In the synthesis of protein by certain bacteria, e.g., *Escherichia coli* (Gross and Gross, 1956), there are some mutants of this organism which require a specific amino acid. If this amino acid is not supplied then the synthesis of both protein and RNA comes to a stop. If the chloramphenicol is given nothing further happens but, if in the next stage a small amount of the required amino acid is added, then RNA synthesis starts up very rapidly without affecting the protein synthesis. Then, if the chloramphenicol is removed from the system, protein synthesis will start again and proceed very rapidly.

It has been mentioned that RNA is present in mitochondria and mitochondria are capable of synthesizing protein. Studies in the author's laboratory by Sheridan have already been mentioned in which it was found that, in the liver cells of guinea pigs that were subjected to scurvy, the mitochondria were surrounded by layers and layers of ergastoplasmic membranes, almost as if these were being synthesized on the surface of the mitochondria. In fact they are so closely applied to the surface of the mitochondria in some cases that it is very difficult to distinguish them from the mitochondrial membrane. Whether this means, in fact, that there is synthesis of this membrane going on or whether it means that the membrane has become closely apposed to the mitochondria, because the energy derived from mitochondria is required in the process of protein synthesis as carried out by the endoplasmic reticulum, one cannot tell. There is also a possibility that, since in scurvy at least some forms of protein synthesis are depressed, the mitochondria may be synthesizing additional endoplasmic reticular membranes in an attempt to compensate for the depressing effects of the vitamin C deficiency. Mitochondria are enormously increased in numbers in scurvy and also in plain starvation.

It is of interest that the RNA particles of the endoplasmic reticulum like those of the nucleoli measure approximately 150 A across, and hence uniformity of size as well as uniformity of structure may play a part in protein synthesis. We have already mentioned that the synthesis of RNA which itself determines synthesis of protein is under the control of the DNA, i.e., the nucleus, although we mentioned earlier that there are both DNA and RNA in the mitochondria. This we know, for example, because a Mendelian gene controls the sequence of amino acids in human hemoglobin as shown by the work on sickle-cell anemia and furthermore, spermatozoa transfer only DNA and no RNA unless there is some in the mito- chondrial sheaths of the middle piece.

What is the significance of the protein part of the RNA particle we do not know, presumably it is structural though it may be en- zymic in nature. Crick has commented, "The RNA forms the tem- plate and the protein supports and protects the RNA." Crick has suggested that the ribonucleoprotein particle is an open structure like a sponge and possibly molecules of appropriate size can diffuse in and out of it.

Recent studies by Dr. Harry Harris of Oxford University, using hybrid tissue-culture cells, have added more information about the role of the nucleolus in information transfer in the cell. Dr. Harris believes that this flow of information is regulated by the nucleoli. When messenger RNA is produced by the DNA of the nucleus, it apparently accumulates in the nucleoli where it becomes incorporated into protein to form a ribosome and then passes from the nucleus into the cytoplasm. The studies of Ficq and others concerning the synthesis of proteins by nucleoli probably indicate that protein for attaching to RNA to form a ribosome is synthesized there. This would naturally take priority over protein synthesis in the cytoplasm since the latter could not occur until the ribosomes were synthesized in the nucleolus and passed to the cytoplasm. If the nucleoli of cells are destroyed by a fine beam of ultraviolet light, RNA does not pass out from the nuclei to the cytoplasm.

The first work on DNA and messenger RNA was done on bacterial cells which do not have the typical nucleus and nucleolus of higher (eukaryote) cells. In bacterial cells it was found that as RNA is produced on the DNA template it diffuses to protein particles in the cytoplasm where it carries out its instructions for protein synthesis. The suggested role for the nucleolus helps to tidy up a lot of unexplained findings, but a new revolutionary idea that for the time being at least, will complicate our interpretation of protein metabolism has just been published by Dr. Eugene Bell of the Massachusetts Institute of Technology. His work indicates that the following may be the model for protein synthesis in the cell. He proposes a messenger DNA instead of RNA, which he calls informa- tional DNA. Pieces of this 1-DNA pass to the cytoplasm where, if they are associated with protein, they form 1-somes. According to Dr. Bell, it is this DNA that serves as the template for the synthesis of RNA and he believes that transcription of RNA may actually start before the DNA is finally packaged in its 1-some, perhaps even while the DNA is still associated with the outer nuclear membrane. He suggests that the 1-DNA forms a "ternary complex with the ribosomes attached to the DNA that is synthesized from the 1-DNA. If this is correct, then translation might occur simultaneously with transcription (or virtually so) or follow it, but only after an 1-some—ribosome complex arises."

How can the previous findings of the formation of ribosomes in the nucleolus and their passage to the cytoplasm be reconciled to this? It was mentioned earlier in this chapter that there is nucleolus-associated DNA and perhaps this is actually the 1-DNA of Dr. Bell and perhaps it produces RNA in the nucleolus; the 1-some—ribosome complex may be formed right there and passed out into the cytoplasm. However, further work will, sometime in the future, undoubtedly reconcile this revolutionary suggestion with existing findings.

NUCLEOCYTOPLASMIC RELATIONSHIPS

There are many problems in the relationship between the nucleus and the cytoplasm and their mutual interreactions. In the first place we should consider such reactions in unicellular organisms. Full details of this subject should be obtained by reading Brachet's "Biochemical Cytology." See also L. Goldstein's articles in "Cytology of Cell Physiology" (3rd ed., Academic Press, 1964). Only a summary of the very interesting facts presented by them can be given here.

We know that the nucleus is largely composed of DNA, RNA, histone, and a number of proteins. There is a high concentration of RNA in the nucleoli that is very labile, "Messenger RNA," which carries information from the nucleus to the cytoplasm, is formed in association with the DNA of the nucleus. Both histochemically and biochemically it has been demonstrated that nuclei contain extremely little of the oxidative enzymes (cytochrome oxidase and succinic dehydrogenase) that are so characteristic of mitochondria. So that presumably the chemical events that take place in the nucleus are those that are predominantly anaerobic in nature. Nuclei contain glycolytic enzymes and also an enzyme that synthesizes

DPN, using nicotinamide, nucleotide, and ATP. There are also present a number of enzymes that are concerned in purine and nucleoside metabolism. These occur in greater quantity in the nucleus than in the cytoplasm, at least as far as liver cells are concerned. It has been suggested by various authors, including Brachet, that these findings are in agreement with the fact that the nucleus could be the site of nucleotide, coenzyme, and nucleic acid synthesis. Many of the experiments that have endeavored to show some relationship between the various cellular components have been carried out on homogenates or by mixing of, say, nuclear fractions and mitochondrial fractions and so on, but these are likely to give a very misleading idea of what in fact does happen in the living cell. How important this is, is demonstrated by very interesting experiments carried out by de Fonbrune in 1939. He transplanted nuclei from one ameba to another and showed that the nuclei control the characteristic streaming of the cytoplasm in each case so that the type of streaming of the cytoplasm of one ameba is transferred to another species of ameba when the nucleus of the first cell

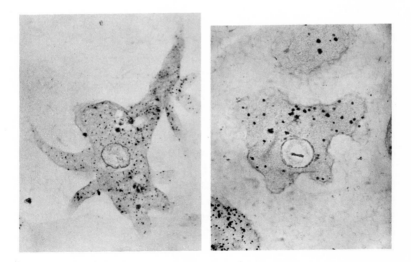

FIG. 111 This illustration demonstrates that a protein (ribonuclease) can penetrate the cell membranes of living cells. Left, a normal ameba stained to demonstrate RNA. Right, RNA stain in another ameba that has been treated with ribonuclease. Note decrease in number of granules in ameba on right. (From Brachet, "Biochemical Cytology," Academic Press, New York, 1957.)

is pushed into the second. The point of these experiments, however, is that in the process of transfer the cell walls of the two amebas must be in extremely close contact with each other and the nuclei are then simply pushed through the two cytoplasmic walls by means of a blunt glass probe. If, however, the nucleus is pushed out into the medium and pushed into the other cell, it loses its ability to divide and, in fact, many of its activities seem to stop. So one can only guess at what happens to the nuclei and mitochondria and possibly other parts of the cells when the latter are homogenized, washed in sucrose, spun down in centrifuges, drawn up in pipets, squirted out of the pipets, mixed up with various reagents, and so on. It is pretty certain that the cell organelles being used in this way are in a different state from those that exist in the living cell. It is interesting that Cutter (1955), and his colleagues have demonstrated that coconut milk contains a number of nuclei that swim freely in it and that presumably come from the endosperm cell. These nuclei, which are in a physiological liquid that the coconut supplies, are still not capable of mitosis and are capable only of degenerative division when they are transplanted back into endosperm cells.

Studies have been carried out on the ameba and on the unicellular alga, *Acetabularia,* which involve sectioning the cytoplasm so that one-half contains a nucleus and the other does not. The groups of workers concerned with these studies are Hammerling and his colleagues, Mazia, and Danielli and co-workers. See articles by J. Brachet, J. Hammerling and colleagues, J. F. Danielli and colleagues and other workers in "The relationship between nucleus and cytoplasm" [*Exptl. Cell Res. Suppl.* **6** (1959)]. If the nucleus is removed from an ameba, the animal quickly loses its motility and ceases to put out pseudopodia, rounds up, and becomes spherical. It is of interest, however, that if the nucleus is removed from ciliates, the part that has no nucleus still retains ciliary action. It cannot engulf living organisms but, if both the nucleate half of the ameba and the nonnucleate half are kept starving, the nucleate half dies within about three weeks and the half without a nucleus in two weeks so the former has about 50% longer survival resulting, presumably, from the presence of the nucleus. If the nucleus is grafted into a fragment of cytoplasm of the ameba that has no nucleus and if the animal has only been in this condition for two or three days, then there is a very dramatic reintroduction of pseudopod formation and characteristic ameboid motility. However, if the half has been without a nucleus for about a week or longer then the introduction of a nucleus does not have the same rejuvenating effect, and it is apparent that irreversible changes have taken place in the cytoplasm.

Danielli has expressed the opinion that particularly in these amoebas and possibly in cells of higher animals the nucleus is responsible for the *type* of macromolecules which are produced whereas the cytoplasm plays the part of organization of these macromolecules into what he describes as functional units. On the other hand, there are probably also mutual interreactions between the cytoplasm and the nucleus; it is not just all one-way control. (Fig. 111 shows penetration of ribonuclease into ameba and its effects on RNA.)

Hammerling has carried out a number of extremely interesting experiments on *Acetabularia* (see Fig. 112 for diagram of structure of this alga), and we will take this opportunity of referring here to some of them. *Acetabularia,* a unicellular alga, normally has a stalk that contains some chloroplasts and a series of rhizoids, and in one of the larger rhizoids the nucleus is situated. At an appropriate period, the stalk forms a cap, which looks rather like

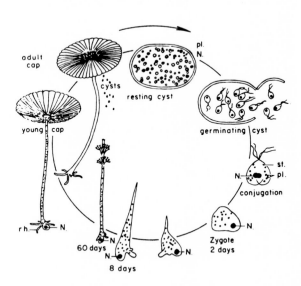

FIG. 112   Life cycle of *Acetabularia mediterranea.* (From Brachet, "Biochemical Cytology," Academic Press, New York, 1957.)

that of a mushroom, and produces cysts that germinate and undergo a sexual union to form a zygote that reproduces the stalk and the rhizoids again. If a second rhizoid containing a nucleus is grafted onto the stalk of an acetabularian, it will form two of these caps instead of one. Here is an example of the effect of the nucleus on the cytoplasm. In the reverse experiment, Hammerling showed that, if a cap is removed from the acetabularia just before the nucleus goes into division, then nuclear division will stop and will not occur until a new cap has been formed. Presumably something necessary for the nucleus to divide is secreted by the cap, and, if the cap is continually removed, the process of mitosis can be delayed indefinitely. On the other hand, if the nucleate part of a rhizoid is grafted onto a plant that already has a nucleus, then a cap nuclear division can be produced within two weeks instead of what would be the normal time of about two months. In *Spirogyra* it was shown 50 years ago that, if the nucleus is removed, the nonnuclear parts of the cell survived for quite a long time, carried out photosynthesis, formation of plastids and fats, and production of tannic acid; the protoplasm continued to stream and there was even an increase in length. Thus a good deal of activity can go on in the cytoplasm even without the nucleus and yet in the end the cell processes do stop and the cell cannot continue to survive. True regeneration cannot occur without the presence of a nucleus.

Loeb, in 1899, originally thought that the function of the nucleus was that of a center for cell oxidation, but the fact that the nucleus contains very little oxidative enzyme activity by biochemical studies and gives little or no reaction for these histochemically is evidence that this suggestion was not correct.

It is an interesting fact, in view of what has been said of the synthesis of coenzymes a little earlier, that, in 1925, E. B. Wilson had stated that the nucleus might be a storehouse of enzymes or of substances that activated cytoplasmic enzymes and that these substances may be concerned with synthesis as well as with destructive processes. It was suggested also, in 1892, by Verworn that the nucleus was the main synthetic center of the cell. This suggestion was also supported by Caspersson in the light of his various studies, and he thought that the nucleus was the principal center for protein formation in the cell. Mazia in 1952 suggested that the function of the nucleus is really that of a replacement of products of cell activities and in support of this he points out that removal of the nucleus is not followed by effects that take place immediately but instead they take place over a period of time. For example, the nucleus produces enzymes, if the nucleus is removed then the cytoplasm will continue to function with the enzymes it has until they drop below a functional level—this may take place at different rates for different enzymes. Mazia has also suggested that as an alternative the nucleus might produce what could be described as a cytoplasmic unit that would be comparable to a plasmagene, presumably this would be messenger RNA. It would play a part in cytoplasmic synthesis and would have to be replaced all the time by the nucleus if the cytoplasm was to be maintained normally. As soon as these cytoplasmic units were exhausted then the cytoplasm would not carry on its normal activities. Thus, although the nucleus may not be the center of oxidative processes in the cell, it may influence these processes and we should consider the evidence for and against this.

Figure 113 demonstrates the influence of the nucleus over RNA production.

Shapiro, in 1935, by the process of micrurgy cut sea-urchin eggs into two pieces, one half contained the nucleus and one did not, and he found that the oxygen consumption was much higher in the fragment that contained no nucleus. This is as one would expect from our knowledge that the cytoplasm contains mitochondria

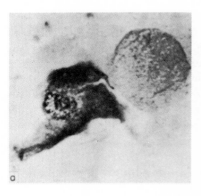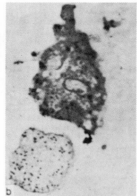

FIG. 113 Two amebas with severed portions. Note that anucleate portions show loss of basophilia, i.e., loss of RNA. (From Brachet, "Biochemical Cytology," Academic Press, New York, 1957.)

and that the mitochondria contain most of the respiratory enzymes of the cell. Brachet has also studied the oxidative processes in *Ameba proteus* (1955) and he found that the removal of a nucleus had very little effect on the rate of respiration and, in fact, there was no change for at least seven days, but after that time the cytoplasmic fragments began to undergo cytolysis and the respiratory rate then fell off. The drop in the latter is understandable since the maintenance of the proper structural organization of the mitochondria and probably the related cytoplasm is necessary for the respiratory rate to be maintained. In the case of *Acetabularia* a similar result has been obtained, however, some studies that were carried out by Whiteley (1956) on the protozoan, *Stentor*, are of interest in this connection. He too severed his protozoan into two portions, one of which contained the nucleus and one did not and showed that in the fragment without a nucleus there was a drop in the consumption of oxygen and no regeneration, however, in the fragments that had a nucleus, regeneration occurred after 24 hr and there was a corresponding increase in O_2 consumption during this period. Eventually the respiratory rate came back to normal.

These results suggest that the macronucleus has some control, perhaps not over the actual respiration but over the synthesis of fresh respiratory equipment, in other words, of fresh mitochondria. It is in fact known that certain of the mitochondrial enzyme proteins are synthesized under the influence of messenger RNA from the nucleus although some others can be synthesized by mitochondrial RNA and DNA. The results with the amebas and with *Acetabularia,* as Brachet says, completely invalidate Loeb's older conception that the nucleus was the center of oxidation processes in the cell. In fact, it is quite striking that the oxygen uptake is unaffected for quite a long time after the nucleus is removed from cells—in the case of the ameba this is as long as ten days, in the case of *Acetabularia* as long as three months—and one cannot conclude from this that the nucleus once it has formed respiratory machinery exerts any significant control over its functioning. On the contrary, it appears that the nucleus is extremely dependent on the cytoplasm, and we have noticed a little earlier how important it is for the nucleus to be kept in contact with cytoplasm. Its removal from cytoplasm for even a short time renders it incapable of carrying on its processes of division.

A relationship between the nucleus and cytoplasmic components is suggested by the results of tissue culture workers who have frequently described the migration of mitochondria within the cytoplasm of the cell so that they come into contact with the nuclear membrane, remain there for some time and finally disengage themselves. The nucleolus has often been observed to move through the nucleus and touch the nuclear membrane on the inside. It has been mentioned that usually it is met there by mitochondria. The cyclic movements of nucleolus and mitochondria in spinal ganglion cells noted by Tewari and the present author have also been already mentioned. In these cases it is possible that the nucleolus is discharging RNA or DPN through the nuclear membrane or, in the case where the mitochondrion is associated with the cell membrane, ATP is perhaps passed inward to the nucleolus.

Although the nucleus does not appear to have any special relationship to respiratory activity in the cytoplasm, it does have other effects of quite a fundamental nature. This has been very well demonstrated by experiments that have been carried out on amebas by Brachet and his colleagues and have given extremely interesting results. They showed that if amebas are cut into nucleate

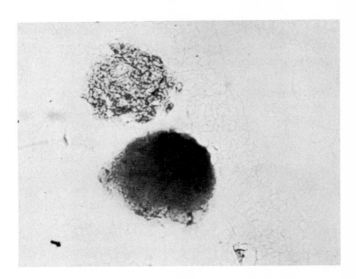

FIG. 114 Two halves or amebas. Dark glycogen staining of lower (anucleate) portion indicates failure to metabolize glycogen. (From Bracket, "Biochemical Cytology," Academic Press, New York, 1957.)

and nonnucleate parts, although the oxygen consumption is not significanty affected by this process, there is a disappearance of the ability to utilize glycogen in the nonnucleate part. In other words, the glycogen stores are scarcely affected (see Fig. 114), but, on the other hand, the protein tends to be used up so that the nucleus appears to control the process of glycogenolysis and also maintains the integrity of the protein of the cell. It is of interest that it takes a period of about three days before the carbohydrate breakdown stops and protein catabolism becomes increased. It is obvious, therefore, that the nucleus exercises very profound control over fundamental processes in the metabolism of the cytoplasm. Although there is a lag period before changes in the glycolytic system and protein metabolism occur in the nonnucleate part of the ameba, changes in phosphate metabolism take place almost instantaneously. It was originally demonstrated by Mazia and Hirschfield in 1950 that the uptake of radioactive phosphate by the nonnucleated part of the ameba was substantially less than that of the part containing the nucleus. Brachet and his colleagues repeated this experiment. First they placed whole amebas in radioactive phosphate for some time and then separated the amebas into nucleated and nonnucleated parts and demonstrated that there was no particular uptake of the radiophosphorus in the nucleated part as compared with the non-nucleated part. Then they presectioned amebas and put the sectioned portions, nonnucleated and nucleated, into the radiophosphate and demonstrated that there was a rapid uptake of radioactive phosphorus in the nucleated portion and a great reduction of uptake in the nonnucleated portion, and, in fact, after a period of about six days there was 30 times more radioactive phosphorus in the former portion. This indicates, therefore, that the phosphate metabolism is profoundly affected in the absence of the nucleus. This result suggests that possibly the synthesis of ATP and of nucleotides are affected in the half without the nucleus. However, it was demonstrated by Brachet that the nonnucleated part actually had more ATP and not less. On the other hand, what might be happening here is that the utilization of ATP has been reduced or greatly slowed down in the nonnucleated portion compared with the nucleated portion or that the ATP was used up in the nucleated portion in the production of nucleotides. This is particularly likely in view of the fact that phosphate uptake is reduced in the absence of a nucleus. This reduction brings us to the very interesting conclusion that we are dealing here with a situation in which the nucleus is responsible for the coupling of respiratory activity with oxidative phosphorylation, and it is quite possible, therefore, that when the nucleus is removed respiration is uncoupled from oxidative phosphorylation and this latter process comes to a dead stop even though the respiratory activity is unaffected. How the nucleus performs this function, whether it is by the production of a coenzyme, as Brachet has suggested, or not is difficult to say. A number of explanations have been put forward of this effect, but one of the most plausible is that, since by far the greater percentage of the DPN-synthesizing enzyme is present in the nucleus, its removal deprives the cytoplasm of most of this critical coenzyme. Brachet makes this point too in his discussion of the problem. It was Hogeboom and Schneider who showed in 1952 that the DPN-synthesizing enzyme is concentrated largely in the nuclei of liver cells and that the nuclei themselves contain considerable quantities of DPN. It has also been found to be present in fairly high concentrations in the *nucleoli* of the eggs of starfish. Since the nucleus controls the production of DPN, it is obvious that there would be a drop in DPN in the fraction of cytoplasm in which it is not present. Now DPN is essential for glycogenolysis and this would explain the holdup in this process in nonnucleated fragments. DPN, however, is also required for respiration, but the DPN concerned with respiration is not present in the cytoplasm but is conserved in the mitochondria and could conceivably not be affected by the process of removal of the nucleus. The DPN that is present in the cytoplasm and is concerned with glycogenolysis might well be susceptible to breakdown by DPNases in the absence of a nucleus. The lack of nuclear produced DPN would be responsible for this chain of reactions grinding to a stop. Attempts to prove this hypothesis, which was enunciated by Brachet, were made by two authors, one of whom found that the DPN content in the nonnucleated portion of an ameba decreased by more than half in a period of 75 hr after enucleation; the other author reported no change in the DPN level for six days. As Brachet says, one will need to wait until this conflict is resolved before it is possible to assert or confirm this theory of the relationship of the nucleus to cytoplasmic respiration and oxidative phosphorylation.

When similar experiments are carried out on *Acetabularia,* the story does not seem quite the same. There appears to be very little difference in uptake of radiophosphorus between *Acetabularia* with nucleus or without nucleus. However, the situation is complicated by the fact that this alga carries out the process of photosynthesis and that photosynthetic processes are unaffected by the removal of the nucleus, and in fact the uptake of radioactive phosphate is concerned with photosynthesis. With *Acetabularia* it appears then that the nucleus does not exercise the same control as in animal cells.

One of the things that happens when an ameba is enucleated is that the cytoplasm changes from a fibrillar appearance under the microscope to a granular one. At the same time the basophilia decreases in the nonnucleated portion and continues to decrease for some days after the removal of the nucleus. By the fifth day there is practically no basophilia in the fragment that does not contain the nucleus. All this suggests that there is breakdown of the endoplasmic reticulum, especially the ergastoplasmic part; the loss of basophilia presumably means loss of the RNA particles which were associated with these membranes.

A very ingenious and interesting experiment was conducted by Goldstein and Plaut in 1955 and described by Brachet. These two authors cultivated organisms known as Tetrahymena, which they had immersed in radioactive phosphate solution and then amebas were permitted to feed on them. Presumably then the radioactive phosphorus that was liberated into the cytoplasm of the amebas was absorbed into the nucleus and possibly converted into RNA. After the digestion of the *Tetrahymena* the nucleus from an ameba that had ingested these radioactive animals was then grafted either into an ameba that already had a nucleus or into an enucleated ameba. In the case of the latter, within a relatively short period of time, 12 hr or so, radioactive RNA was found to be present in the cytoplasm. That this was RNA was demonstrated by the fact that, after treatment of this organism with ribonuclease, no radioactive particles could be shown by autoradiographic techniques to be present in the cytoplasm.

Another point of interest is that where such a nucleus was poked into an ameba that already had its own nucleus, radioactive particles of RNA soon appeared in the cytoplasm but these particles did not pass back into the original nucleus, which remained free of radioactivity during the whole of the experiment. This indicates that the passage of RNA is probably one way only, that is, from the nucleus into the cytoplasm.

Brachet has pointed out, however, that Goldstein and Plaut have themselves made the point that the radioactive material that is discharged from the nucleus into the cytoplasm is not necessarily RNA that has been formed in the nucleus. It may, in fact, be a precursor or it may be that the radioactive phosphate is taken up in the nucleus and passed back into the cytoplasm as such and then synthesized as RNA in the cytoplasm itself. However, even if this is so, it is of interest that the nucleus is a source of phosphate that is used for RNA synthesis in the cytoplasm or a source of a precursor that is used for this purpose. However, studies over the past ten years have established that a good deal of the cytoplasmic RNA comes from the nucleus. Brachet makes the point that, if the RNA metabolism is affected in the enucleated ameba, there should be some inhibition in the synthesis of protein since there are many pieces of evidence that link these two processes together as has been described earlier. Mazia and Prescott in 1955 found that the amount of methionine labeled with radioactive sulfur (^{35}S), which is incorporated into a nonnucleated ameba fragment is two and a half times less than the amount incorporated in the fragment with the nucleus. This difference, however, does not occur until three days after the ameba has been severed in two. This provides some evidence that there is a link between protein synthesis and RNA synthesis and that in this indirect way, by control over the cytoplasmic RNA, the nucleus also controls the protein synthesis in the cell. It should be noted that protein synthesis does not fall to zero in fragments without a nucleus, so there is a residual cytoplasmic synthesis of proteins that persists even although the nucleus is removed from the cytoplasm and this is probably due to the DNA and RNA of the mitochondria.

It is not known whether the nucleus controls the synthesis of all the different types of cytoplasmic protein. We know, however, as mentioned earlier, that synthesis of some mitochondrial protein appears to be independent of the nucleus. The nucleus does, however, control the synthesis of some other mitochondrial proteins. Brachet has studied changes in a number of different enzymes in the nonnucleate halves of amebas and has demonstrated that various types of enzymes are affected in different ways by the nucleus. (Enzymes are, of course, proteins so the study of the enzymes gives an indication of the synthesis of this particular type of protein.) Protease, enolase, and adenosine triphosphatase are enzymes that appear to undergo very little or no change once the nucleus is removed. Amylase, on the other hand, seems to increase slightly in activity then drops back to normal level so that it is not affected very much. Dipeptidase decreases in the beginning and then remains at a constant lower level, whereas acid phosphatase and esterase are reduced progressively and after a few days cannot be demonstrated as being present at all. Thus different enzymes are obviously under nuclear control to different degrees, and it becomes obvious that control of the nucleus over cytoplasmic proteins is a very complex one.

It is also of interest that Danielli and his colleagues in 1955 made hybrid amebas by putting the nucleus of one species into the cytoplasm of another and then prepared antibodies against the two species. They found that the lysis of the hybrid ameba is under the control of the nucleus. This appears to demonstrate that the determination of antigenic specific characters, as Brachet puts it, is under the "nuclear dominance while the morphological and physiological character of the hybrid is under cytoplasmic dominancy." It is of interest that in our earlier discussions in this book on the role of the nucleolus on the synthesis of RNA we mentioned that the dissected out nucleoli of *Acetabularia* exposed to radioactive phosphate had incorporated ^{32}P with extraordinary rapidity into their RNA so this is a very good support of the claim that the nucleolus is a center of RNA synthesis. RNA is of course synthesised elsewhere in the nucleus. It is of interest that in the case of *Acetabularia* the removal of the nucleus actually stimulates RNA synthesis in the cytoplasm, but it is also to be noted that in intact *Acetabularia* the nucleolus is much more active than the cytoplasm in the production of RNA. In the case of "protein synthesis" (absorption of labeled amino acids) in *Acetabularia,* it has been demonstrated that the amino acids appeared in the nucleolus before they appear in the cytoplasm and that once the nucleus is removed, protein synthesis by the nonnucleated portion is actually faster than in the nucleated portion and this "ties in" well with the demonstrated increase of RNA synthesis. It also shows that the nucleus is not essential for the synthesis of protein by the cytoplasm, but further experiments have demonstrated that such anucleate protein synthesis is of a short-term nature (about three weeks) and that for it to be extended over a long period of time the nucleus is necessary. It is of interest that, although the synthesis comes to an end, the turnover (breakdown) of protein continues for quite a number of weeks even if the nucleus is not present.

A striking thing about the intact cell is the metabolic activity of the RNA contained in the nucleolus in comparison with that found in the cytoplasm. Some early work suggested that incorporation of precursors into RNA was faster in the nucleolus than in the cytoplasm, and there is now a good deal of evidence that this is so mostly from autoradiographic studies by Ficq in 1955. By working with the oöcytes of echinoderms and amphibians, she demonstrated that radioactive labeled adenine and orotic acid were incorporated very rapidly into RNA in the nucleolus. In the lampbrush chromosomes of amphibian eggs, adenine could be found to be incorporated very rapidly into the loops of the chromosomes, which are known to contain RNA. There have also been some studies with radioactive phosphate and a number of authors have demonstrated that it too is preferentially absorbed into the nucleolus. All these findings, together with the previous literature, strongly support the conception that the nucleolus in particular is a very active site of synthesis of RNA and that it is the source of at least part of the cytoplasmic RNA.

The incorporation of labeled amino acids into the protein of the nucleus has been subject to some small controversy. Originally it was believed that the rate of incorporation into the nucleus was no more rapid than into the cytoplasm, but this was probably due to the fact that the work was carried out on aqueous homogenates and that nuclei lose a certain amount of their protein in such homogenates. It was not until homogenates were studied by nonaqueous techniques that it was found that labeled amino acids were incorporated into nuclear proteins very rapidly. Brachet points out that the proteins of the nucleus are presumably either the constituents of the chromatin or they represent protein that is localized in the nucleolus, and it is not possible to tell for certain with the homogenate technique just where in the nucleus the proteins are localized.

Ficq and Brachet have demonstrated that in the liver cells of higher animals the nuclear proteins incorporate radioactive amino acids much more actively than cytoplasmic proteins but in the case of the nuclei of the pancreas, the intestine, the lung, heart, muscle, kidneys, spleen, and uterus this is not so.

In developing embryos it has been demonstrated that all the

nuclei show a very much higher incorporation of radioactive amino acids than the cytoplasm and since, as Brachet points out, there is a considerable synthesis of proteins in the nucleus in actively dividing cells, this is just what would be expected.

Brachet has given a general discussion of the possible relationships of the nucleus and the cytoplasm and the readers are referred to his book for full details. However, certain of the more essential conclusions might be repeated here. One of the functions that the nucleus could perform since it is relatively poor in enzymes is the production of coenzymes, and it is possible that some of the enzymes found in the cytoplasm might be under the regulation of the coenzymes secreted by the nucleus. The importance of coenzymes might explain the association of the mitochondria with the nuclear membrane that has been recorded by quite a number of authors. However, we only know for certain that DPN as a coenzyme is made by the nucleus and we do not know for certain what other coenzymes might be produced by it. It is of interest that Tewari and Bourne have produced histochemical evidence that the nucleolus in spinal ganglion neurones synthesize ATPase and glucose-6-phosphatase and passes them first to the body of the nucleus then to the cytoplasm. The striking effect, however, of this association is that the activities of the mitochondria seem to be pretty well independent of the nucleus insofar as cellular oxidation is concerned since the oxidation proceeds whether the nucleus is there or not. However, as we have pointed out earlier, it is an interesting possibility that the nucleus acts as a device for coupling by some means or other oxidative phosphorylation with respiration. The independence or otherwise of the mitochondria from the nucleus is of interest, and Brachet points out as we have done earlier in this discussion that the nucleus may in fact depend upon its supply of energy (ATP) from the mitochondria.

Where, however, the nucleus appears to exert its effect most strongly is on the endoplasmic reticulum, and, since it has been demonstrated that great decrease of basophilia and change in the nature of the cytoplasm occurs when the nucleus is removed from the ameba, it seems possible that the real relationship between the nucleus and the cell lies in the control by the former over the endoplasmic reticulum. Its control here may be dependent on the secretion of RNA granules which become attached to the reticulum as ribosomes. (See Fig. 115 for summary of division of labor in cells.)

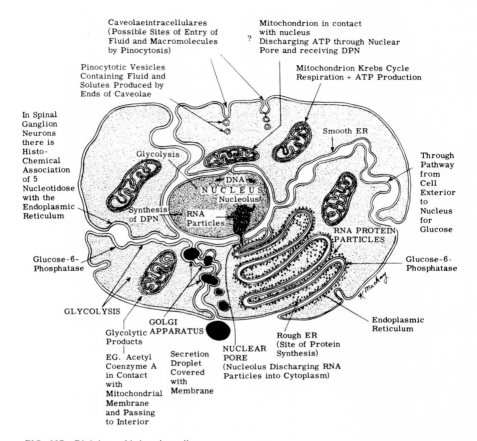

Caveolaeintracellulares
(Possible Sites of Entry of
Fluid and Macromolecules
by Pinocytosis)

Mitochondrion in contact
with nucleus
? Discharging ATP through Nuclear
Pore and receiving DPN

Pinocytotic Vesicles
Containing Fluid and
Solutes Produced by
Ends of Caveolae

Mitochondrion Krebs Cycle
Respiration + ATP Production

In Spinal
Ganglion
Neurons
there is
Histo-
Chemical
Association
of 5
Nucleotidose
with the
Endoplasmic
Reticulum

Glycolysis

Smooth ER

DNA
NUCLEUS
Nucleolus

Through
Pathway
from
Cell
Exterior
to
Nucleus
for
Glucose

Synthesis
of DPN

RNA
Particles

RNA PROTEIN
PARTICLES

Glucose-6-
Phosphatase

Glucose-6-
Phosphatase

GLYCOLYSIS

Endoplasmic
Reticulum

GOLGI
Glycolytic APPARATUS
Products

EG. Acetyl
Coenzyme A
in Contact
with
Mitochondrial
Membrane
and Passing
to Interior

Secretion
Droplet
Covered
with
Membrane

NUCLEAR
PORE
(Nucleolus Discharging RNA
Particles into Cytoplasm)

Rough ER
(Site of Protein
Synthesis)

FIG. 115 Division of labor in cells.

Originally it was suggested by Caspersson that DNA was really
the nucleic acid fundamentally concerned with the synthesis of
complex proteins. More recently it has been suggested that DNA
probably directly synthesizes the proteins of the chromosomes, in
other words it causes a reproduction of the protein portion of the
genes; whether it does this directly is not clear.

There is no doubt that RNA plays an important part in embryological processes and it might be of interest in concluding this chapter
to repeat Brachet's conception of its role. "Since the synthesis
of proteins is primarily a cytoplasmic process and since, as we

know, the RNA present in the ergastoplasmic small granules certainly plays a part in the synthesis, it is tempting to visualize the whole process in the following, entirely hypothetical way. Under the influence of the nuclear genes, specific cytoplasmic RNA's would be synthesized; these RNA's are probably not entirely synthesized in the nucleus, and it is possible that interactions between purely cytoplasmic RNA and RNA of nuclear origin occur. The specific RNA's would in turn organize the synthesis of specific cytoplasmic proteins according to the well-known template hypothesis.''

All the studies that have taken place since Brachet made this statement serve to lend it their support. A summary of cell structure is shown in Fig. 115.

SEVEN
Specialized
cells:
gland
cells,
muscle
fibers,
and
nerve
fibers

UP to date we have been dealing with cells in a rather general way, using a variety of cell types to illustrate points concerning cell physiology. It is proposed now in the last stages of this book to consider two or three specific types of cells. To begin this chapter, we will consider the glandular cell and, since there are a wide variety of both exocrine and endocrine glands in the body, comments will be restricted to the type of secretion process that is demonstrated particularly well either in the salivary glands or the pancreas and particularly we will refer to the pancreas, although some details of the salivary glands will be included as well.

The mechanism of secretion in externally secreting glands has been a subject of study for a good many years, and very detailed work was published on this subject by Nassonow in 1923; Bowen, in a series of publications between 1924 and 1929, fully confirmed and extended Nassonow's work. The general trend of studies by Nassonow and Bowen was the relationship of the Golgi material in particular to the production of secretion droplets. They both demonstrated that the first droplets of secretion that are visible under the optical microscope appeared in the interstices of the Golgi network, that is the number and and size of these droplets increased, the Golgi network became hypertrophied and that when the cell carried its full load of secretion the apparatus appeared to break up and small pieces of it appeared to become attached to some of the secretory granules.

Studies have been made on the pancreatic cells of external secretion (as distinct from the islets of Langerhans) from the point of view of fine structure. The pancreatic cell structure can thus be taken as a general model for the cells of external secreting glands.

Two groups have made a special study of the fine structure of the acinar cells of the pancreas. These two are Ekholm and Edlund (1959), who investigated the human pancreas and Palade and Siekewitz (1958), who studied the guinea pig pancreas. Generally speaking there was little difference between the two types of pancreas. The same cytoplasmic elements—the endoplasmic reticulum with associated ribonucleoprotein granules, the mitochondria, and the zymogen granules were present in both.

In guinea pigs that had been starved, Siekewitz and Palade demonstrated that there was a well-formed endoplasmic reticulum (ergastoplasm) and that the cisternae (spaces between the pairs of membranes) were very small and had special orientation within the cell. There were very few granules in the cisternal spaces. The apical parts of the cell contained a number of zymogen granules and the lumina of the acini were very small and appeared to be occupied by an amorphous material that had a density similar to that of the zymogen granules. This finding agrees with the statement that secretion of zymogen does occur from the pancreas in small amounts even in long starvation. In guinea pigs that had been killed 1 hr after feeding, there were many granules within the cisternae of the endoplasmic reticulum that were greatly distended. The number of zymogen granules found in the apex of the cell was generally less than in the starved animals. The lumina of the acini were distended and contained irregular and ill-defined masses of electron-dense material that was probably discharged zymogen.

Siekewitz and Palade subsequently made a very interesting study of the functional variations in the microsomal fractions in the pancreas of the guinea pig. The microsomal fraction as mentioned before represents the endoplasmic reticulum plus its attached ribonucleic acid particles. Their experiments started from the fact that they observed that, when the microsomal fraction from starved guinea pigs was analyzed for enzyme activity, it was found that the ribonuclease and TAPase (trypsin activatable protease) activities varied a great deal in different preparations and in some cases reached as high as 30% of the activities found in the zymogen fraction. The zymogen fraction is that part of the homogenate of the pancreatic cells that is formed from the granules in the cell that are described as the zymogen granules and that are primarily comprised of digestive enzymes and precursors of enzymes—hence the name "zymogen." Since the zymogen fraction is composed almost exclusively of zymogen granules and since these have a higher total RNAase and TAPase activity than any of the other cell fractions, the authors wondered whether this similarity indicated some functional connection, and they initiated a series of experiments in order to test this out.

In other investigations they had found that in some cases dense granules had been found within the cisternae of the endoplasmic reticulum in electron microscope preparations of the exocrine cells of the pancreas of the guinea pig. These intracisternal granules could be compared, both from the point of view of fine texture that they demonstrated and their density, with the zymogen granules. However, there was a considerable difference in size and the zymogen granules appeared to be outside the endoplasmic reticulum whereas the smaller particles were actually in the cisternae between the double membranes. Further studies have demonstrated that these intracisternal granules are present in large numbers in this position about 1 and 2 hr after starved guinea pigs have been fed but that by 3 to 4 hr they have more or less disappeared. The authors consider that, since it is the endoplasmic reticulum that forms the microsomes in homogenization, perhaps some of the high enzyme activity demonstrated in the microsomal fraction might have been due to the enclosure of some of these granules that were present within the microsomal cavity. They investigated this problem by taking various fractions from starved and fed animals, collecting the glands about 1 hr after feeding, the time when they expected to obtain the highest concentration of granules within the cisternae of the endoplasmic reticulum. They were able to obtain subfractions 1 hr after feeding that consisted largely of cisternal granules together with some detritus derived from the microsomal fraction. They found that the TAPase and the RNAase activities present in this granular fraction were actually higher than those of the microsomes from which they were presumably derived and were sometimes as high as the values for the zymogen fraction as such. A similar

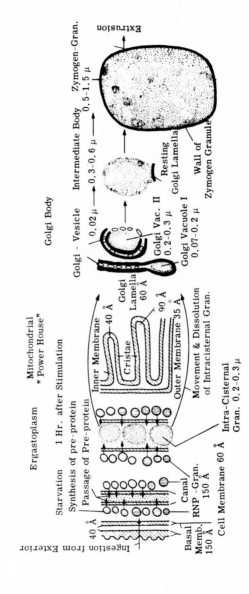

FIG. 116 Hirsch's conception of "production line" secretion by a cell. (From *Naturwiss.* **47**, 25, 1960.)

increase in activity was also described for the enzyme amylase. The granules present in the cisternae were already in a finished state, and were probably destined therefore to become zymogen granules, but their precise position in the secretory circle is not clearly seen at the moment. It is known that in the late stages of the secretory cycle of the pancreas the large vacuoles in the Golgi zone become filled up with a material that has a high electron density, and it is possible that these intracisternal granules feed into the Golgi material and there condense into large granules that are passed to the apical part of the cell. Hirsch has said that in many cases it appears that the granules that are shed into the lumen of the pancreas appear to be enclosed in a structure resembling a membrane of cytoplasmic material. This might be explained in the following way. Assuming that there is a continuous passage from the endoplasmic reticulum lumen through the cisternae in the Golgi apparatus and so to the exterior, it is possible that Siekewitz and Palade's intracisternal granules pass to the Golgi apparatus and there coalesce to form large granules, which then pass away from the Golgi material toward the apical part of the cell along those parts of the endoplasmic reticulum that lead to the exterior of the cell. Since these granules are very big when they leave the Golgi apparatus, they no doubt cause great dilation of the cisternae and it is even possible that as they are pushed out of the cell they tear part of the double membranes from the ergastoplasm and carry this as an investment as they pass out into the exterior. Even if there is no continuous passage it is quite possible that the droplets might be excreted from the cell covered with a submicroscopic skin derived from the endoplasmic reticulum or the Golgi membranes or simply composed of protein secreted on to them by one of these organelles.

Siekewitz and Palade point out that in their preparations, although they found a very high proteolytic RNAase activity in the isolated intracisternal granules, quite considerable proteolytic activity and RNAase activity were also found in the less dense fragments of the microsomes which presumably are composed simply of endoplasmic reticulum and the small RNA particles attached to them. It appears very likely that this activity is localized in the RNA protein particles. It is possible that these individual particles synthesize enzymes, become greatly enlarged, and pass through the membrane of the endoplasmic reticulum into the cisternae. The interrelation between cellular organelles according to Hirsch is shown in Fig. 116.

This represents a scheme for the production of secretion granules by the cell and the joint role of the endoplasmic reticulum, the mitochondria, and the Golgi apparatus in this process. According to this scheme, in the pancreas amino acids become associated with the RNA granules of the endoplasmic reticulum membrane (left of diagram) and are passed as preprotein (? polypeptides) into the cisternae where they grow in size to become the intracisternal granules of Palade and Siekewitz. (This stage is headed "1 hr after stimulation" in the diagram.)

In the text of this book, the previous observations of Hirsch, in which he has seen "secretion" droplets in the pancreatic acinar cell moving through the cytoplasm to the Golgi zone have been mentioned. These droplets are approximately the same size as the intracisternal granules described by Palade and Siekewitz. Hirsch now believes that his moving droplets represent the intracisternal granules passing along the endoplasmic reticulum. The large secretion droplets of gland cells appear in the Golgi apparatus and if the intracisternal granules represent presecretory material their transference from the endoplasmic reticulum to the Golgi material requires some explanation. There is no evidence that the cisternae of the endoplasmic reticulum are continuous with the Golgi material but this possibility must be considered. Hirsch believes that the mitochondria provide the energy required to dissolve the cisternal granules, to pass them through the membranes of the endoplasmic reticulum and through the Golgi membranes inside which they become concentrated into droplets of secretion, growing larger and larger until they are converted into zymogen granules, surrounded by a membrane, and pushed to apex of the cell ready for discharge. See Fig. 117 for an electron micrograph of secretion droplets in a pancreas cell.

This is the morphological story to date as told by the electron and light microscopes. The metabolic story has been studied, using salivary gland material by Junquiera and his colleagues with rats as their experimental animals. Normal cats and rats with ligated salivary ducts were used. The significance of ligating the excretory duct is that it leads to cessation of secretion, disappearance of granules of secretion, decrease in size of the cells and the gland as a whole, and there is a corresponding decrease in activities of the enzymes. The gland does not undergo any degenerative process. Junquiera and his colleagues used these types of glands to study not only the histochemical and biochemical changes in the cells but they tried also to study the morphological changes as demonstrated by the light microscope in an attempt to equate the various activities and morphological elements in the cell with the processes of secretion. These results are given in summary as follows. In the control gland it was found that there were many mitochondria; they were mainly in the form of short rods though some round specimens were present. As a result of ligation they were very greatly reduced in numbers. Millon's histochemical test for proteins demonstrated a strong reaction in the normally secreting gland and only a very weak reaction in the gland that had been ligated. Studies in basophilia (to demonstrate degree of accumulation of RNA) showed a very strong basophilia in the control gland, but in the ligated gland, which was apparently not secreting at all the basophilia was very weak. Biochemical investigation of the ratio of ribonucleoprotein to deoxyribonucleoprotein (RNAP/DNAP) showed this ratio to be 2.18 in cells of the control gland but only 0.84 in the ligated gland, so there was obviously a very profound decrease of RNA by comparison with DNA in the nonsecreting gland. Biochemical studies of protease activity demonstrated that in the control gland it was 560.0 μg phenol/100 mg tissue and in the ligated gland it was 266.0 μg phenol/100 mg tissue. In the mouse, the cathepsin activity was 16.9 μg tyrosine/1 μg DNA protein and in the ligated gland it was not reduced quite as much as some of the other enzymes but was down to 10.3 μg tyrosine/1 μg DNAP. Alkaline phosphatase varied a little between rats and mice. In rats the figure was 2.8 mg phenol/100 mg tissue and in the ligated gland it was 1.7 mg phenol/100 mg tissue, so this was a reduction of something like a third. In mice, curiously enough, in the control gland there was 2.0 mg phenol/100 mg tissue of alkaline phosphatase and in the ligated gland it actually went up to 2.4 mg phenol/100 mg tissue. So it seems that if the alkaline phosphatase is playing a significant part in the secretory processes of rats, it is a specific process and it does not play the same part in mice.

The acid phosphatase results were interesting because in rats the control gland showed 1.8 mg phenol/100 mg tissue activity and in the ligated glands there was no change at all. In mice, however, in the control gland there was 2.0 mg phenol/100 mg tissue activity and in the ligated gland 0.42 mg phenol/100 mg tissue, so that,

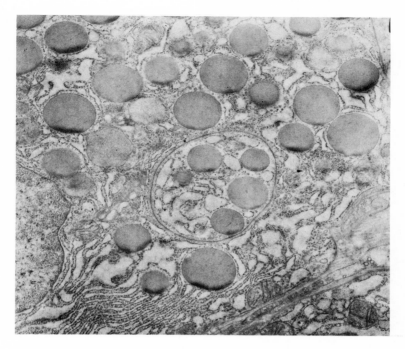

FIG. 117   Electron micrograph of pancreas cell of rhesus monkey. Magnification 21,500×. The edge of the nucleus can be seen on the left. Below it and to the right is well-defined endoplasmic reticulum. Secretion droplets can be seen amid the endoplasmic reticulum in other parts of the cell. In the center of the illustration fine droplets appear to have become encircled by a loop of endoplasmic reticulum.

perhaps, if phosphatase is playing a part in the secretory cycle, then in the rat it is alkaline phosphatase that is involved and acid phosphatase in the mouse.

Succinic dehydrogenase was studied by Thunberg's methylene blue method and there was found to be high activity (50% reduction in 12 min of methylene blue) in the control gland and in the ligated gland there was a 12% reduction in color, which was decreased in mice to less than a quarter. There was not quite as much reduction in rats; using the Warburg apparatus there was a 63% increase of the QO_2 after the addition of succinate and in the ligated glands only 28% increase. So there was a significant reduction in oxidative activities, which was confirmed by the cytochemical indophenol oxidase test for cytochrome oxidase, the control gland cells giving a dark blue and ligated glands only a light blue color with this reagent. The oxygen consumption in the rat cells gave a QO_2 of 4.2 in the control gland and only 1.8 in the ligated gland. The figure for glycolysis and the glycolytic quotient in relation to nitrogen was 3.2 in the control gland and 3.3 in the ligated gland, so there was significant difference in the glycolytic rate. The ATP and ADP combined figures was 13.7 μg/1 μg of DNA protein in the control gland and in the ligated gland it was only 3.2 μg/1 μg of DNA protein. Creatine phosphorus was 8.5 μg/1 μg DNAP* in the control gland, and this was down to 2.7 μg/1 μg DNAP in the ligated gland. Inorganic phosphorus dropped from 44.8 μg/1 μg of DNAP in the control to 18.6 μg/1 μg of DNAP in the ligated gland. Pyruvate utilization similarly dropped by about two-thirds, and thus it appears from these figures that not only is there a drop in respiratory activity of the cells when they are not secreting, but there is also a drop, as one might expect, in oxidative phosphorylation. It is of interest, however, that glycolysis is maintained at a constant rate in both ligated and nonligated cells, and, as Junquiera and his colleagues point out, it appears that the energy derived from glycolysis itself is used by the cell largely for its basic needs and not for its specialized function as a secreting cell. It appears that the ATP, ADP, and phosphocreatine are really the compounds that are the sources of energy for the secretory process in the cell since there is such a spectacular decrease in these following ligation. It is of interest that the activity of cathepsin in the secretion of the salivary glands seems to be controlled *in vivo* by the male sex hormones. There is a parallel between cathepsin activity and protein synthesis, and the drop in activity of cathepsin in the cells of ligated glands may be significant from this point of view.

Junquiera and Hirsch have also described the light-microscope changes that take place in cells of ligated glands.

The Golgi apparatus is undoubtedly related to the processes of secretion if the early work that we have mentioned before can be taken at its face value. Junquiera and Hirsch summarized the evidence for the participation of Golgi material in the process of secretion as follows. The Golgi apparatus changes in size and structure very considerably during the secretory cycle. Droplets appear in the substance of the Golgi material and gradually appear to be

---

* DNAP = DNA phosphorus.

transformed into secretory granules. This fact has been observed not only in fixed and stained preparations but also in the living pancreas, for instance, by Hirsch in 1932. In 1939 he suggested that the Golgi bodies are the sites in the cell where cytoplasmic products congregate and become formed into zymogen granules. There are no biochemical data, as has been pointed out both by the present author and by Hirsch, that the Golgi bodies are actually the site of protein synthesis since they contain no RNA and do not appear to contain oxidative enzymes or other enzymes concerned with energy production, although in some cells such as the absorptive cells of the gut they do contain phosphatase, ATPase (heat stable), alkaline phosphatase, acid phosphatase, and so on.

In 1954, Sjöstrand and Hanson demonstrated by ultrastructural studies that there was an intimate relationship between the Golgi vacuoles, the zymogen granules, the ground substance of the Golgi apparatus and the alpha cytomembranes (endoplasmic reticulum): "There are all transition stages observed between, on the one hand, bodies with a pronounced elongated form and with the most direct topographic relations to the Golgi membranes and the Golgi granules embedded in the ground substance on the other. The impression when observing these pictures is that they show snapshots of the conversion of membranes into granules and vice versa. The small granules seem to coalesce to bigger granules which gradually gain the size, form and opacity of the zymogen granules." This work appears to confirm the conception of the Golgi material as a condensing or aggregating region of the cell and reminds one of the statement of Kirkman and Severinghaus in 1938. "A great deal of work strongly suggests that the Golgi apparatus neither synthesizes secretory substances nor is transformed directly into them but acts as a condensation membrane for the concentration, into droplets or granules, of products elaborated elsewhere and diffused into the cytoplasm. These elaborated products may be lipoids, yolk, bile constituents, enzymes, hormones or almost any other form of substance."

We do not know, of course, to what extent these synthetic processes are controlled by the mitochondria. There are papers in the literature suggesting that the mitochondria actually produce the zymogen granulus in the pancreas, and it seems very likely that in parts of the endoplasmic reticulum the mitochondria come in very intimate contact with the nucleoprotein granules on the outside of the membrane and that they supply the energy for the synthetic processes taking place in these particles. At any rate the particles of enzyme protein that are produced pass into the spaces and presumably undergo some benefit from contact with the Golgi material and possibly even from contact with the nucleus. One could conceive of them as all passing into the cisternal space that surrounds the nucleus, but to what extent the metabolic activity of the nucleus might affect any enzyme protein passing through this space one cannot say for certain. Perhaps the passage of the granules to the Golgi region serves not only as a mechanism whereby the granules are aggregated to a large zymogen granule but possibly something is done to prepare the enzymes for activity upon secretion. Also one should consider the possibility that vitamin C, despite the known defects of localization in the technique, may occur, as the present author has claimed, in the Golgi apparatus at times of great synthetic activity and there is a possibility that in passing through into an area saturated with a reducing substance, oxidation of the contents of these granules prior to their excretion may be prevented.

So we can see a division of labor very clearly in the process of secretion in the gland cell but the interesting thing, as we have been emphasizing right through this book, is that we are not dealing with an isolated division of labor but an interdigitating division so that each job fits in very well with the job done by another part of the cell.

THE STRIATED MUSCLE FIBER

The muscle fiber is composed of four fundamental constituents: (1) the sarcolemma that is the equivalent of the cell membrane; (2) the fibrils that represent the structural elements responsible for contraction; (3) the sarcosomes that are really the mitochondria and contribute the supply of energy for muscular contraction; (4) the sarcoplasm, which is the ground cytoplasmic substance in which the other structures of the muscle fiber are embedded; and (5) the sarcoplasmic reticulum. In addition, there is a specialized region of the fiber called the motor end plate at the point where the motor nerve fiber makes junction with the sarcolemmal membrane, which is responsible for initiating the contraction of the fiber. For the moment we will discuss very briefly the structure of each of these elements, and we will include the description of the structure of the caveolae intracellulares with an account of the structure of the sarcolemma.

Muscle fibers themselves have three main shapes. They may be cylindrical with ends that are conical, they may be spindle-shaped with extremities tapering off very finely, or they may be conical with one end long and attenuated and a broad base at the other. This depends, of course, on their position in the muscle itself. The first of these types of fibers usually runs the whole length of the muscle, the second type is usually situated within the main part or belly of the muscle, the third type is usually attached to a tendon at one end and the other end terminating somewhere in the interior of the muscle itself.

The shape of a transverse section of muscle is oval or spherical when it is fresh but it shrinks considerably following fixation, and one usually finds rather angular cross sections of the fibers. Voluntary muscle fibers vary from about 10 to 100 μ in diameter. Presumably this difference in size is associated in some way with the amount of work the muscle has to perform. There are in any single muscle, fibers of many different diameters. Fibers vary very much in length and may in some cases extend only a few millimeters and in others, fibers in excess of 34 cm in length have been seen.

Schwann in 1839 and Bowman in 1840 described a thin membrane that surrounded the voluntary muscle fiber. It was described by Schwann as the cell membrane of the fiber and this was later agreed to by Bowman. It can be demonstrated by placing a muscle fiber in fresh water or by causing a sudden coagulation of the contents which then contract. The membrane or sarcolemma, as demonstrated in this way, seems to be a pale, colorless and apparently, under the optical microscope, structureless membrane, and it is semipermeable since if placed in water it swells by imbibition and if subsequently placed in concentrated sugar solution this water is removed. It was thought to be about 0.1 μ across and to have a slight infolding at each Z band. Others have shown that the sarcolemma is more complicated, i.e, under the electron microscope there are two dark osmiophilic lines separated by a light osmiophilic line, which together are approximately 300 Å across, each of the lines in this triple structure being about 100 Å thick; also there is a thick layer of dense material about 500 Å thick, which extends inward from the inner of the dark lines of the membrane.

Under the light microscope the early workers had described

the sarcolemmal membrane as being about 0.1 μ thick, which is in fact 1000 Å. We will see that, if we add the 500-Å-thick complex to the three lines that themselves total about 300 Å, we get a figure of about 800 Å, which is pretty close to the 0.1-μ size of the sarcolemma described by the optical microscopists. As Robertson points out we are faced, of course, with the decision as to which of these many structures can be regarded as the true membrane. What we are in fact dealing with here is really a membrane complex rather than a single membrane and this, strictly speaking, applies to the membranes of all other cells as well. Every now and again the sarcolemma is penetrated by holes that are the ends of the caveolae intracellulares. Some of these little openings or caveolae expand into vesicles and in some cases both the vesicles and the caveolae themselves may extend in a complicated fashion, branching and ramifying over and in between the myofibrils. It has been suggested by Bennett and also by Palade that such caveolae are not a standardized feature of the structure of the sarcolemma of the muscle fiber, in general, but that they represent a mechanism whereby ions or macromolecules such as carbohydrates or proteins can obtain access into the interior of the fibers without violating the osmotic properties of the membrane. There do not appear to be pores in the sarcolemmal membrane so far as present investigations have shown, but Bennett has suggested in his chapter on the fine structure of muscle in the present author's "Structure and Function of Muscle" (Academic Press, 1960), that various substances (e.g., proteins and carbohydrates) would be bound to the outer surface of the sarcolemma. According to the nature of the binding sites there, the membrane would then infold, form a caveola, and this would seal off to form a vesicle and the vesicle would then pass into the interior and might then be destroyed so that the substances become free in the cytoplasm. This a pinocytotic type of interpretation of these caveolae and the possibility exists, of course, that this type of activity may occur in many types of cells.

Compounds could be passed out of the muscle fiber in a similar way, that is, by becoming attached to the internal surface of the plasma membrane, which would then surround it and open up to form a caveola and the contents would become liberated to the exterior. Thus large macromolecules could get through the membrane without actually piercing it.

It is of interest that in connection with this theory Whatman and Mostyn (1955) noted droplets of fat that had been stained with fat dyes in the process of passing through the sarcolemma of cardiac muscle fibers. Quite a number of cells and perhaps all cells appear to have attached on the outside some sort of polysaccharide coating to the plasma membrane. Such a coating occurs on the surface of mammalian erythrocytes where it is extremely thin, and it has been pointed out that this may represent the surface coating that causes the ABO agglutination reaction.

Plant cells are, of course, expert in putting polysaccharides around the outside of their cell membranes, and bacteria consistently do this too. Outside the sarcolemma there also appears to be a cloud of material, approximately 500 Å thick, which is possibly polysaccharide material. It is of interest that there is a positive periodic acid–Schiff (PAS) reaction in the sarcolemma, and this may be the cause of the reaction. This significance of the polysaccharide complex may be its ability to bind various molecules, which are to be taken into the interior of the muscle membrane.

In the region of the motor end plates there is considerable modification of the sarcolemma, and Robertson in 1956 and Couteaux in a series of papers have published detailed accounts of this. The sarcolemma in the region of the motor end plate is very much thicker, denser, and more folded particularly into a series of troughs that penetrate deeply into the sarcoplasm of the muscle fiber. The external polysaccharide-like material just described is present everywhere and separates the actual endings of the neural components of the motor end plate from the sarcolemmal membrane itself.

The motor end plate as demonstrated by the cholinesterase technique is shown by Fig. 118.

The sarcoplasm of the muscle fiber was originally described by Spidel in 1939 as being in a gelled condition; this can change to a sol if there is any damage to the muscle.

The sarcoplasm not only surrounds the nucleus and is obvious in that part of the fiber (see Fig. 119), but it also penetrates and surrounds and fills in the spaces between the myofibrils themselves and between the myofibrils and sarcolemma. The relative amounts of sarcoplasm in the fibers varies very considerably, and it is of interest that Golarz and Bourne have demonstrated great activity for the enzyme acetyl phosphatase in the sarcoplasm of human mus-

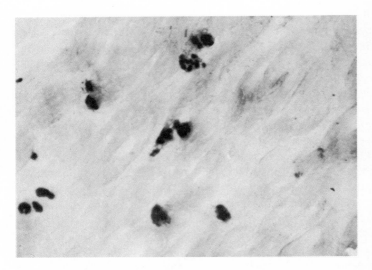

FIG. 118 Cholinesterase in human neuromuscular junctions (motor end plates). (Preparation and photograph by E. Beckett.)

cle. The sarcoplasm also contains most of the soluble proteins of the muscle fiber. Under the electron microscope it is less dense than the other components of the fiber. There are also a number of small granules scattered about the sarcoplasm that show a great range in size, something between ten and a couple of hundred angstroms across. They are distributed without any particular pattern, and in some regions of the sarcoplasm there do not appear to be any at all. They appear to be more abundant in certain areas of the sarcoplasm in heart muscle fibers, and, according to Fawcett and Selby (1958) they may, in fact, be composed of glycogen. Such areas give a strong PAS reaction, and, if the section is treated with amylase prior to application of the PAS reaction, the reaction disappears. This is strong evidence that these granules are, in fact, glycogen.

Beckett and Bourne have demonstrated that PAS positive material was present, which was dispersed irregularly in human muscle fibers, sometimes as irregularly distributed aggregates and sometimes as fine granules. To some extent the difference in size of these granules depends on the changes that take place following death. A biopsy specimen, for instance, shows only small granules

FIG. 119 Electron micrograph of frog sartorius muscle fiber, nucleus, and mitochondria massed in sarcoplasm. (Preparation and photograph by R. Q. Cox, Dept. of Anatomy, Emory Univ.)

and aggregations but by 12 hr the aggregates have reached their optimum size and after 48 hr autolytic procedures have removed them altogether. Beckett and Bourne have demonstrated that the amount of stainable material with the PAS technique in normal muscles is extremely variable and that there is no correlation between the amount present and the anatomical site of the muscle concerned. Also in muscular and neuromusclar disorders the amount of stainable material is variable and inconstant and shows no particular relationship with the neuromuscular diseases that were studied. For details see the article on this subject in "Structure and Function of Muscle" [(G. H. Bourne, ed.), Vol. 3. Academic Press, 1961].

It is of interest that Dempsey and his colleagues and Beckett and Bourne were unable to remove all this PAS positive material with either diastase or saliva. This suggests that we are dealing either with glycogen, which is linked to and protected by protein in some way or with a mucopolysaccharide type of substance, or possibly even aldehyde groups are being produced by the PAS reaction by lipids or fat. Beckett and Bourne also applied the McManus PAS technique to formalin fixed, frozen sections in conjunction with the Sudan black test for fats. These were made both on rat and human muscle. By this frozen section method it was found that the muscle fiber was stained in a regular diffuse positive PAS background staining and there were also droplets of the material giving a strikingly intense positive PAS reaction. These droplets are often considerably increased in pathological muscles and appear to accumulate at points of mechanical damage in muscle fibers. These strongly positive droplets do not appear to be glycogen. Paraffin embedding of the muscle, however, removes these droplets. It is possible because of these facts that they may be a lipid of some sort. They are not, however, sudanophilic (thus are probably not fat) and do not give a Schultz reaction for cholesterol.

Many muscle fibers contain droplets of fat in the sarcoplasm. These vary very considerably in individual muscle fibers, in individual muscles, and in individual muscles as a whole. Some muscles have more fat than other muscles. This was first described as long ago as 1841 by Henle and subsequently investigated by Kölliker (1846), and others. Fat was recorded in human muscle as long ago as 1889. Beckett and Bourne using the Sudan black technique in human muscle found that fat could occur in droplets of very variable size,

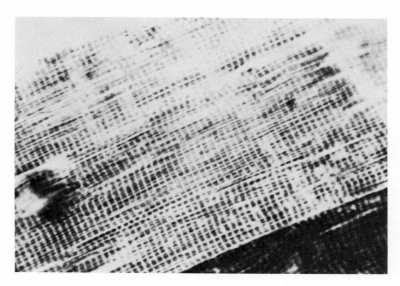

FIG. 120 Cross striations in skeletal muscle fibers of rat. The division of the fibers into longitudinal fibrils can be seen. (Preparation by E. Beckett; photograph by the present author.)

distributed at random, but that a few were situated at the poles of the nuclei. Also there was some staining in the regions of the cross striations with Sudan black.

Some muscle fibers have a red color and some have a white color. Red and white muscles are well known among mammals, and in animals such as the rabbit and guinea pig and also in the turkey both red and white muscles occur separately, but generally speaking, particularly in humans, muscles are a mixture of both red and white fibers. The red color of the fiber depends upon the fact that it contains myoglobin or muscle hemoglobin. The red fibers contract more slowly and remain contracted for a longer period than the white muscles. There are more of them in those muscles that are concerned with posture. Generally speaking too, the red muscle fibers contain more sarcoplasm than the white and they also appear to contain more fat, which presumably functions as a reserve supply of energy for the muscle fiber. (See Fig. 121A–D for enzymes in muscles.)

Ribonucleic acid granules have not been identified in the sacroplasm of adult muscle. An interesting new development in the struc-

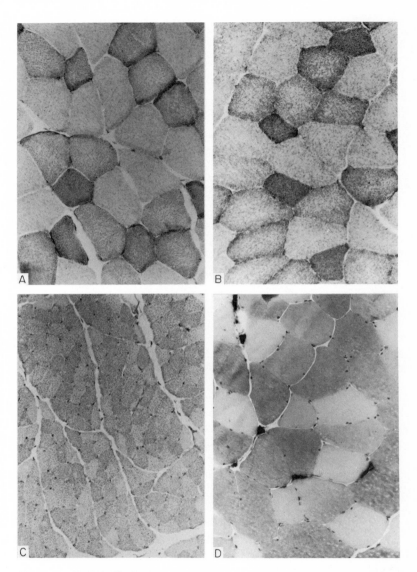

FIG. 121 A. DPNH—Diphosphopyridine nucleotide diaphorase, skeletal muscle, *Macaca mulatta* (×192). B. LDH—Lactic dehydrogenase, skeletal muscle, *Macaca mulatta* (×192). C. Hexokinase, skeletal muscle, *Macaca mulatta* (×75). D. Adenosine triphosphatase, skeletal muscle, *Macaca mulatta* (×192). (Photographs by M. Nelly Golarz.)

ture of muscle fibers is the delineation of the sarcoplasmic reticulum described by Bennett in his chapter in "Structure and Function of Muscle" [(G. H. Bourne, ed.), Vol. 1. Academic Press, 1960]. As Bennett points out, for more than 100 years papers have reported that there appears to be some sort of a network or series of tubules associated with muscle fibers. Retzius in 1881, had described a sarcoplasmic network and Kölliker had referred to a network in 1888, however, the most detailed treatment was given by Veratty in 1902. He described a "reticular apparatus" in muscle; this is a confusing term since the Golgi apparatus had often been referred to in these terms. Bennett and Porter in 1953, found this component in electron-microscope pictures of muscle and described it as a sarcoplasmic reticulum. Sjöstrand and his colleagues subsequently referred to them as sarcotubules. It is of interest that the work on this reticulum that was described and understood very well by these early workers appeared to be completely ignored in later books of histology where one finds scarcely any reference at all to what now appears to be a very important structure in the muscle fibers.

Porter and Palade made a very detailed study of the sarcoplasmic reticulum with the electron microscope (see Figs. 122 and 123). The reticulum appears to be arranged in the form of alternating sets of tubes that anastomose with each other and surround the fiber like a bracelet. The anastomosing of the tubules gives the general impression of a lacelike type of structure. One set of these tubules has a plane of symmetry at the Z band and extends on each side of it to about the junction between the A and I bands; alternating between these is a series of tubules which is in the plane of the M bands. It should be emphasized that these structures which are being described are characteristic both of skeletal muscle and cardiac muscle.

The transverse element, which has its plane of symmetry parallel with the Z band, has been shown to be a separate structure from the sarcoplasmic reticulum, although it comes in very close contact with it at the Z band. This transverse element is called the "T system." The T-system membrane is continuous with the sarcolemma. Professor A. F. Huxley of the University of Cambridge in England has shown, using a microelectrode, that electrical stimulation of the Z band causes a contraction of the fiber, presumably due to the stimulation of the T system and its closely associated sarcoplasmic reticulum.

It is now believed that the following course of events involving this system takes place in muscle contraction:

Stimulation of the sarcolemma by the nerve impulse via the motor end plate passes into the T system causing it and the sarcoplasmic reticulum to release Ca ions. These ions trigger enzyme reactions associated with myosin to cause a loss of phosphate from ATP with liberation of energy. This is used for the shortening of the

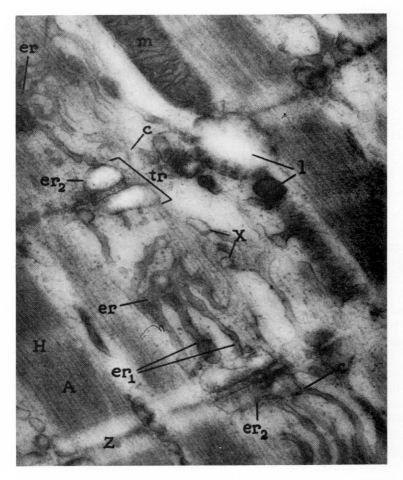

FIG. 122 Sarcoplasmic reticulum. (From Porter and Palade, *J. Biophys. Biochem. Cytol.* **3**, 269, 1957.)

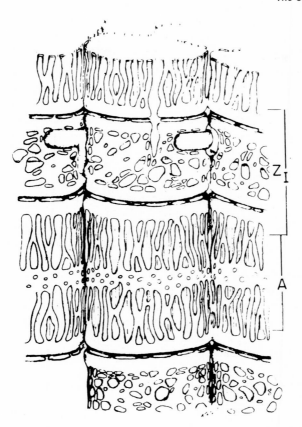

FIG. 123 A schematic summary and interpretation of observations on the
sarcoplasmic reticulum of rat sartorius muscle. A single myofibril occupies
the center of the image with portions of adjacent myofibrils surrounding it. A
complete sarcomere with associated reticulum is pictured as part of the
central fibril. Sarcosomes, shown in less than normal number, occupy well-
defined positions relative to the Z line. They are pictured with their long axes
oriented circumferentially with respect to the fibril and covered or not by the
close reticulum of the I-band region. Approximately 32,000×. (Legend and
figure from Porter and Palade, *J. Biophys. Biochem. Cytol.* **3,** 269, 1957.)

myofibrils (contraction). The Ca ions are then reabsorbed back onto
the T system and sarcoplasmic reticulum membranes and the myofibrils relax and ATP becomes regenerated. It is believed that some
ATP can be synthesized by the sarcoplasmic reticulum in addition
to that produced by the mitochondria so that instantaneous energy

can be provided for muscles that work rapidly in bursts. Such muscles have a well-developed sarcoplasmic reticulum, whereas heart muscle, which works steadily and continuously, has a poorly developed sarcoplasmic reticulum but very well developed sarcosomes.

Characteristic Golgi apparatus is also found in the sarcoplasm of muscle close to the nucleus, as one might expect. Although we have referred to a striated muscle fiber as a cell, probably it is really a number of cells since any one striated muscle fiber may contain a number of nuclei and in a very long fiber, which measures several centimeters in length, some hundreds of nuclei may be present which may multiply by amitosis. They lie at the surface of the fibers (except in cardiac muscle where they are in the center) just under the sarcolemma surrounded by sarcoplasm. The nuclei are generally oval in shape, with their long axes parallel with the long axis of the fiber, they are approximately 8–10 μ long and may be extended as much as 17 μ. They are hard to see in unstained fresh muscle fibers but come out very well in fixed and stained preparations. In muscle diseases and in degenerating muscle, they appear to migrate into the center of the fiber and one can see long strings of muscle nuclei aligned end to end along the center of the fiber. It is of interest that in embryonic muscle fibers the nuclei occupy a middle rather than a hypolemmal (close to the sarcolemma) position.

The fine structure of muscle nuclei does not differ from that of other cells. The nuclei are surrounded by two unit membranes analogous to the membranes of the endoplasmic reticulum of other cells. In fact, Bennett thought the outer membrane of these pairs may be continuous with the cytomembranes that are found in the perinuclear cytoplasm or even continuous with those of the sarcoplasmic reticulum, in which case the arrangement of the sarcoplasmic structure is fundamentally the same in muscle fibers as in other cells. The membrane of the muscle nucleus also contains a number of pores.

The fibrils of the muscle fibers are the contractile elements. They appear to extend for the entire length of the fiber and they are approximately 1–3 μ in diameter. Muscle fibers under the light microscope show a well-defined cross striation (see Fig. 120) and this, of course, is visible in the fibrils; and, it is of interest that the individual fibrils are aligned so that the appropriate segments of their bands coincide with each other. Between the fibrils are

FIG. 124 Striated flight muscle of blowfly (*Calliphora*). Electron micrograph 17,300×. Three fibrils and large sarcosomes characteristic of flight muscles can be seen. The pattern in the central fiber is due to the fact that the thick filaments pass in and out of the plane of section. (From Huxley and Hanson, Abstr., *European Congr. Electron Microscopy, 1st*, p. 202.)

the sarcosomes or muscle mitochondria, to which we will return in a little while. (See Fig. 124.)

It is interesting to note that the fibrils consist simply of orientated protein molecules and do not have a surface membrane; therefore the substances in solution in the cytoplasm can have a direct influence on the protein of the fibrils. They have unrestricted access to each other.

The transverse striations of the fibrils of the muscle fiber are due to a difference in density (see Fig. 125). These cross striations form a pattern and each pattern repeats along the whole of the length of the fiber and each repeating part of the pattern is known as a "sarcomere." In vertebrate muscle each sarcomere is about 2 to 3 μ in length. The fibril itself is constructed of filaments that are made of protein, which lie parallel to the long axis of the fiber and overlap each other. There are two kinds of these filaments, one being about twice the diameter of the other, and they are also different in length. These two types alternate along the length of the fibril and they overlap at parts and interdigitate with each other. The regions where the thick filaments are located are known as the "A" bands, these are extremely dense and are anisotropic under the polarizing microscope. Where the thin filaments are present there is less density and there is less birefringence, these are known as the I or isotropic bands. There is a band that appears across the middle of these which is called a Z membrane or band.

At the center of the A or anisotropic band, thick filaments are present alone but at both ends of the A band there is interdigitation of the thin and the thick filaments and this is the densest part of the fibril. The part of the A band where the thick filaments occur alone is called the H band and is naturally not as dense as the rest of the A band. The Z membrane is due to the presence of a band of amorphous material which occupies the spaces between the filaments at the midpoint of the area where the thin filaments are located (the I band). In the case of the A bands, thickening of the thick filaments themselves about the middle of the band causes them apparently to be cut across by what is described as the M line or as the M strip by some authors.

The filaments that compose the fibrils of the muscle fibers are made up of protein and there appear to be three proteins concerned: myosin, actin, and tropomyosin. There is more myosin than any other protein; it constitutes approximately 54% of the total protein of the

FIG 125 Electron micrograph of frog sartorius muscle. Note myosin (thick) and actin (thin) filaments interdigitating. Note cross linking between filaments. (Preparation and photograph by R. Q. Cox, Dept. of Anatomy, Emory Univ.)

fibrils. There is only about 11% of tropomysin but about 20 to 25% of actin. This leaves a deficiency about 10% in total protein, and it is possible, according to Huxley and Hanson in their chapter in the "Structure and Function of Muscle" [(G. H. Bourne, ed.), Vol. 1, Academic Press, 1960], that the figures for the various proteins are a little too low particularly for tropomyosin and probably for actin too.

It appears that myosin is the protein of the thick filaments that constitute the A bands of the fibers. Actin, on the other hand, appears to be the protein that makes up the thin filaments of the I bands. Tropomyosin also appears to be present where the actin is situated. *In vitro* actin and myosin form a complex called actomyosin, and this can be persuaded to contract under the influence of ATP just as the muscle fiber contracts under appropriate stimulation *in vivo*. Each filament of myosin appears to contain 425 molecules. If myosin is treated with trypsin, it splits into two types of meromyosins—one that is called light and the other heavy meromyosin. These meromyosins apparently occur as distinct units in the myosin molecule. Huxley and Hanson suggest that the backbone of a filament of myosin is made up of light meromyosin units that are longitudinally arranged and staggered, and the heavy meromyosin units, which are attached to the light meromyosins and project from the filaments, these projections can be seen under the electron microscope (see Figs. 125 and 128). Individual myosin molecules have a "head" and a "tail." Each molecule is 1500–1700 Å long. The "head" is about one tenth the length of the whole molecule and has a diameter of about 40 Å—the "tail" is half this thickness and occupies the rest of the length of the molecule. The molecular weight of a myosin molecule is very nearly half a million. The heavy meromyosin mentioned above contains the heads of the molecules and parts of the tails, whereas the light meromyosin fraction contains tails only, and it is of interest that these tails lack ATPase activity, which appears to be concentrated in the heads. The light meromyosin also will not bind with actin. X-ray diffraction studies have shown that the light meromyosin is made up of two polpeptide chains wound around each other in a spiral.

The number of molecules in each filament of actin can also be calculated and they amount to approximately 600. There is some evidence that the actin filaments contain tropomyosin and, according to Huxley and Hanson, there are about 1.7 molecules of actin to

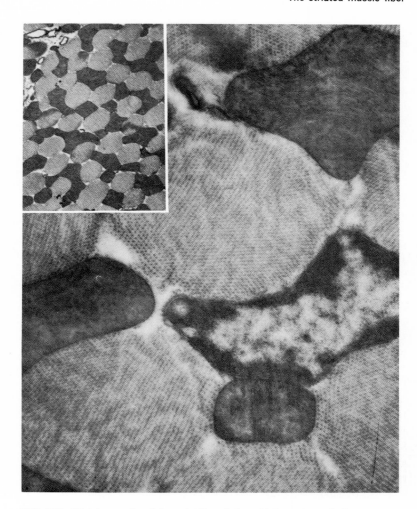

FIG. 126 Flight muscle of insect "Tenebrio." Note extremely large sarcosomes (mitochondria). These organelles are as big as the muscle fibers. (Preparations by David Smith, photograph by Keith Parker. Permission of Phillips Co., Eindhoven, Netherlands.)

1 of tropomyosin. The way these two proteins are linked in the filaments is not known.

A single monomer of "G" actin (many monomers connected together form a polymer—e.g., amino acids are monomers—proteins, polymers; nucleotides are monomers—DNA and RNA are polymers) has a molecular weight of 70,000, much lighter than myosin. Each molecule of actin has a molecule of ATP bound to it and one molecule of "bound" calcium. The actin molecule is spherical in shape measuring about 60 Å across. It is also composed of two strands twisted around each other.

Tropomyosin is a fiber about 400 Å long. It has a molecular weight of 54,000 and is composed of two polypeptides twisted around each other.

Huxley and Hanson believe that the actin molecules are exposed at the surface of the filament so that they may form actomyosin links with the neighboring myosin filaments where they interdigitate. The difficult thing to explain is the fact that a muscle can contract very strongly without significant change in the length of its filaments. There are a number of experiments that have been carried out to elucidate this point. However, it appears that what probably happens is that the shortening of the muscle fiber takes place by a sliding of the actin filaments in between the myosin filaments of the A band; they already interdigitate at their ends and the actin filaments simply slide further in (see Fig. 128). In the noncontracting muscle, the little bumps that have been mentioned before (projections on the heavy meromyosin filaments) come in contact with the actin filaments at the regions of interdigitation and these lock to form an actomyosin complex. Now this happens when no ATP is present, but when ATP is present these locks break apart and the actin filaments can slide up into the myosin filaments. When ATP is dephosphorylated then the locks will form again, and it is possible that in contraction the actin filaments slip into the myosin filaments point by point, there being a locking and releasing corresponding to the production and dephosphorylation of ATP. Then, when all the ATP formed is dephosphorylated, the bands lock again. As soon as ATP is re-formed, then the locks break and actin filaments can slide out from the A bands again and the muscle becomes extended. Thus ATP is necessary both for contraction and relaxation of muscle. It is a very interesting fact that the enzyme ATPase, which is responsible for the dephosphorylation of ATP, is

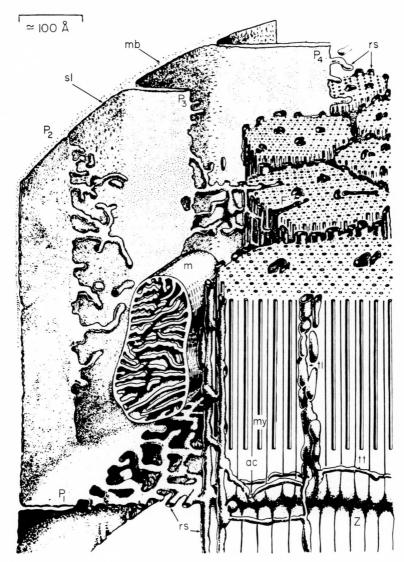

FIG. 127 Three-dimensional reconstruction of muscle of *Cyclops*. (From L. T. Threadgold, "Ultrastructure of the Animal Cell," Pergamon Press, 1967.) ac, actin filament; m, mitochondrion (sarcosome); my, myosin filament; sr, sarcoplasmic reticulum; mb, basement membrane; sl, sarcoplasmic membrane, longitudinal sarcoplasmic reticulum, transverse sarcoplasmic reticulum; P, transverse process.

actually the myosin molecule itself. This is a very interesting example of a structural molecule that is also performing a metabolic function as an enzyme.

One should mention here just briefly one or two of the important chemical reactions that take place in muscle and play a part in the contraction cycle. We have seen that the sliding of the actin filaments into the myosin filaments is dependent on the periodic formation and dephosphorylation of ATP. The re-formation of ATP is a result of the famous Lohman reaction: creatine phosphate + ADP = creatine + ATP. The ATP, as has been mentioned before, is a high-energy phosphate and its dephosphorylation results in the liberation of energy that can be used for the mechanical movements necessary by the molecules in the contraction cycle. We do not propose to give full details here of the chemical reactions that take place in muscles in contraction. Anyone interested in this subject can consult the article "Biochemistry of Muscular Action," by Dorothy Needham in "Structure and Function of Muscle" [(G. H. Bourne, ed.), Vol. 2, Academic Press, 1960]. Szent-Gyorgyi, in the same publication, has emphasized that some molecular contraction must accompany the molecular sliding in muscle contraction.

It has been mentioned earlier that the liberation of Ca ions from the "T system" and sarcoplasmic reticulum seems to initiate the contraction process. It is of interest that, if purified actomyosin is extracted from muscle, its ATPase activity is unaffected by the presence or absence of calcium. However, Dr. Ebashi of Japan showed that, if a protein fraction removed from muscle during the purification of actomyosin were added back to it, the calcium sensitivity returned. This protein was found to consist of two parts: (1) tropomyosin B and (2) a new protein called "troponin." Calcium ions were found to exert their effect through "troponin" in intact muscle.

We might note that the rephosphorylation of ATP requires energy to reconstitute a high-energy bond and this energy can be provided by oxidative phosphorylation produced as a result of coupling respiration with such a process. There are elements in the muscle that are capable of doing this and these elements are the muscle mitochondria, the sarcosomes, which are strategically placed to perform this function. We will say a word now about their structure and position.

Electron-microscope studies of the sarcosomes demonstrate

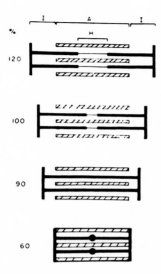

FIG. 128 Diagram showing H. E. Huxley's view of the molecular basis of contraction of muscle. In this theory shortening of the muscle takes place by the sliding of thick filaments, myosin (black) in between the thin filaments, actin (cross hatched). (From Huxley, *Sci. Am.* p. 67, November 1958.)

that they are identical in their fine structure with that of mitochondria. For example, they have a double membrane, and also internal folds of the inner membrane to form double membrane cristae. They are located in the muscle fibers, at least in the case of heart muscle, exactly opposite the A bands. Slater, in his chapter on sarcosomes in "Structure and Function of Muscle" points out that Holmgren had originally classified muscles as falling into two types that depended on the type of distribution of the granules (sarcosomes), and he believed that the two classes corresponded to the two physiological types of muscles. Muscles that had granules at the level of the I bands of the muscle fibrils were those which acted only intermittently, such as the skeletal muscles of most vertebrates, and those which had granules at the level of the A bands were required for continuous activity and in this group would come the flight muscles of birds and insects and, of course, the heart muscles of vertebrates (see Fig. 126).

It is of interest that, for instance, in the flight muscles of the humming bird and of the hovering insects where there is a phenomenal rate of contraction cycles per second (80 per sec in the hummingbird), the mitochondria are relatively enormous in comparison with any other structural elements in the fiber, indicating the high rate of oxygen consumption that must be necessary when the fibers are contracting at such high speeds. Also interesting to note is that sarcosomes, like the sarcoplasmic reticulum, were discovered a good many years ago, in the 19th century in fact. Regaud in 1909 had claimed that they were, indeed, identical with mitochondria because they showed similar staining reactions. Yet, in histology books published since then and up to recent times, sarcosomes have been largely ignored in accounts of the histological structure of muscle. Thus, there are two curious facts about the history of muscle structure—the sarcoplasmic reticulum and the sarcosomes were both discovered more than half a century ago and then dropped out completely from the histological picture and from the textbooks.

Not only do the sarcosomes resemble the mitochondria of other cells in staining reactions and in their fine structure, but they also contain the same series of respiratory enzymes, i.e., those enzymes concerned with the Krebs cycle and cytochrome system found in mitochondria. It is of interest that in red muscle fibers, where, as has been mentioned, the red color is due to myoglobin, there are more mitochondria than in white fibers and that the myoglobin assists in the rapid transfer of oxygen from the blood to the respiratory enzymes found in the mitochondria.

The chemical reactions that take place in muscle can be stated as follows:

$$C_6H_{12}O_6 + 2\ DPN^+ + 2\ ADP + 2\ P\ (inorganic) \rightarrow$$
$$2\ CH_3 \cdot CO \cdot COOH + 2\ DPNH + 2\ H + 2\ ATP \tag{I}$$
$$2\ CH_3 \cdot CO \cdot COOH + 5\ O_2 + 30\ ADP + 30\ P\ (inorganic) \rightarrow$$
$$6\ CO_2 + 4\ H_2O + 30\ ATP \tag{II}$$
$$2\ DPNH + 2\ H^+ + O_2 + 6\ ADP + 6\ P\ (inorganic) \rightarrow$$
$$2\ DPN^+ + 2\ H_2O + 6\ ATP \tag{III}$$

Reaction (I) appears to be restricted to the sarcoplasm, in nonmuscle cells it is in the cytoplasm; reactions (II) and (III) take place in the sarcosomes as they do in the mitochondria. If we sum up those three reactions,

$$C_6H_{12}O_6 + 6\ O_2 + 38\ ADP + 38\ P\ (inorganic) \rightarrow 6\ CO_2 + 6\ H_2O + 38\ ATP$$

we see that the sum of the activities of these processes in the muscle is the synthesis of 38 molecules of ATP for each time the cycle occurs. The function of the sarcosomes in the muscle fibers appears to be the provision of ATP for the functioning of the contractile elements. The myofibrils have no membrane, the sarcosomes are virtually in contact with the myofibrils, and the ATP can diffuse almost instantaneously into the sarcofibrillar filament. If an inadequate supply of oxygen is available or in muscle with an insufficient amount of sarcosomes, the DPNH and the pyruvic acid that are formed as a result of reaction (I) produce lactic acid. This reaction is

$$CH_3 \cdot CO \cdot COOH + DPNH + H^+ \rightarrow CH_3CHOH \cdot COOH + DPN^+$$

It is also of interest that sarcosomes can bring about oxidation of fatty acids, and in the oxidation of stearic acid as much as 147 molecules of ATP are formed with each reaction. Also sarcosomes, similarly to mitochondria and other cell organelles, may become unstable if ATP is not present, and, according to Slater, this suggests that some energy provided by the ATP is necessary to stabilize the structure of these organelles. It is believed that ATP is also used to some extent to maintain a difference in the concentration of ions as between the sarcosomes and the rest of the cytoplasm. Here then, in the muscle fiber we have probably a division of labor as well defined as one could find in any cell in the body. There is a special part of the cell, the contractile part, which is exclusively used for this purpose, not only that but we have the contractile protein acting as an enzyme that can trigger off the whole process. The ATP necessary for the process is provided by the oxidative phosphorylations coupled with the respiration of the mitochondria, which are themselves in close physical contact with the fibrils. The changes in ionic polarization of the membranes that are initiated by the motor endplate and that pass along the sarcolemma can be carried into the fibrils of the muscle fiber by means of the sarcoplasmic endoreticulum. Furthermore, the sarcolemma appears to be adapted to undergo pinocytosis and can take macromolecules directly into the sarcoplasm. This is the perfect example of division of labor in cells, and yet it should be pointed out again how much these individual labors depend on and are integrated with each other. A summary of muscle structure is given in Fig. 127.

THE NERVE FIBER

In conclusion, we will consider the nerve cell and the nerve fiber. The nerve fiber itself is not a cell but a prolongation of a cell. It is a process of a neuron that is adapted for carrying nervous impulses. The functioning of the nerve fiber is a little more mysterious at the moment than is, for instance, the muscle fiber, since we have little in the way of moving parts to investigate and a great deal of the information that is available is theoretical. Most of the studies on the physiology of the nerve fiber are devoted to the movements of ions, particularly sodium and potassium ions, in and out of the nerve-fiber membrane since it is believed that it is the polarization and depolarization of the membrane produced by these ions that constitutes the passage of a nerve impulse along the fiber. The nerve fiber originates from the neuron, from a conical area known as the cone of origin. The nerve cell itself contains a well-defined nucleus and nucleolus. There are numerous Nissl bodies (see Fig. 129).

Among the characteristic structures classically described in the nerve cell are (1) neurofibrils, (2) Nissl bodies (also called tigroid bodies because of their striped appearance, which suggested a resemblance to the coat of a tiger to earlier investigators), (3) Golgi apparatus, (4) mitochondria, (5) fats, phospholipids, and (6) pigments.

Neurofibrils have been claimed to be present in all nerve cells and in the axon, but there is some doubt as to whether they exist as structural elements in the living cells although up to the beginning of the century they had been regarded as the basis of nerve-impulse propagation. There is evidence from microdissection that in the living nerve cells of invertebrates they exist as discrete threadlike structures which are interwoven with each other and appear to be more viscous than the surrounding neuroplasm. In fixed and stained preparations, the neurofibrils are extremely fine and it is not surprising, therefore, that they are hard to see in the living cell. Fibrous structures have, however, been seen in the cytoplasm of nerve cells in tissue cultures and, if fresh nervous tissue is treated with salts such as $CaCl_2$ or various fixatives, the neurofibrils rapidly come into view. One view is that the neurofibril is a rodlet sol, that it is composed of orientated molecular particles that are not, in fact, sufficiently strongly attached to each other to form a definite fibril,

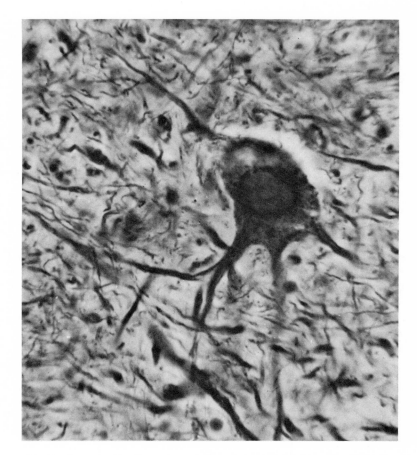

FIG. 129 Photomicrograph of a typical neuron showing processes. Most of those shown are dendrites.

but that a variety of agents may cause these rodlets to precipitate as fibrils.

It seems to be fairly well accepted now that neurofibrillae are a reality. In general they are about 50 to 100 A in diameter. Although they form a network through the body of the cell, there are a number of thick bundles of filaments that are oriented towards the cone of origin of the axon. It has been estimated that in at least 10% of the nerve-cell protein is organized in the form of filaments or fibrillae.

Electron-microscope preparations also show Nissl bodies in nerve cells. They may be stained by basic aniline dyes, and in specially fixed nervous material, histochemical tests have shown that they contain phosphoric acid and iron and they also appear to contain ribonucleic acid (RNA). In certain acute pathological conditions the Nissl bodies may disappear. This is a process known as chromatolysis and, if there is injury to the axon, the appropriate nerve cells show central chromatolysis, that is, the dissolution of the Nissl bodies surrounding the nucleus. There is also a loss of Nissl material from the central neurones of birds in long transmigratory flights. Under ultraviolet light Nissl bodies have the appearance of a flocculant precipitate, and it has been shown that they can be displaced in the cell prior to fixation by ultracentrifugation. It is now known that Nissl substance is composed of masses of endoplasmic reticulum loaded with RNA particles (ribosomes), which serve as a center for protein synthesis for the nerve cell. Because of its large size and the extent of its processes, nerve cells have a high rate of protein production, including many enzymes.

Classic Golgi preparations show that the Golgi apparatus exists as a network close to and wholly or partly surrounding the nucleus, in fact, it was in nerve cells, as we have stated earlier, that Golgi first discovered the apparatus that is named after him.

The form of the Golgi apparatus in neurones is not very different from that of other cells except for its tendency to extend right around the nucleus and to, in some cases, ramify extensively through the cytoplasm. The membranes are arranged in cisternae, and these may enclose spaces that vary from 100 to 700 Å across. The latter are vesicular in shape and are usually connected to the membranes that enclose the narrower spaces.

Mitochondria are fairly obvious in properly stained nerve cells and in cell bodies they tend to be spherical but in the dendrites they are more elongated while in the axon they may appear as long filaments. Near the nucleus there may be an accumulation of fatty-like droplets that may be Golgi material; in a similar position two types of pigment may also be found, e.g., in the *substantia nigra* of the midbrain a black-to-red melanin pigment occurs in this site in the cell. Many neurons show the presence of yellow droplets of lipochrome pigment. This is known as "lipofuscin" and

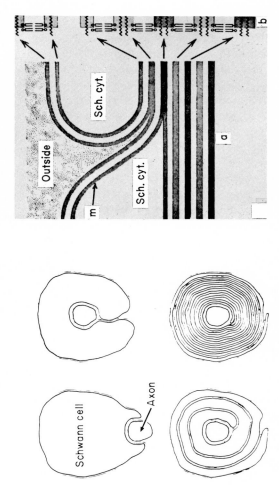

FIG. 130 (left) Schematic representation of the membrane theory of the origin of nerve myelin (after Geren). Four stages in the engulfment of the axon by the Schwann cell and the wrapping of many double layers that, after condensation, form the compact myelin. (Figure and legend from Schmitt, *Rev. Mod. Phys.* **31**, 455, 1959.)

FIG. 131 (right) This diagram illustrates the relationships of the enfolded Schwann cell membrane in a mesaxon (or meso) of an adult myelinated fiber. The two cell membranes come together along their outside surfaces to form the outermost myelin lamella. The inside surfaces of the mesaxon loops are in apposition at the major dense lines in compact myelin. (a) Three complete repeating units of myelin. (b) The molecular structure of the outer two repeating lamessase. At the top the molecular structure of the Schwann cell membrane is indicated. The molecular diagrams are based on Finean's model. The stippling superimposed on the molecular model indicates the densities that are observed after permanganate fixation. (Figure and legend from Schmitt, *Rev. Mod. Phys.* **31**, 455, 1959.)

increases progressively in the cells with age and in the neurons of senescent persons it may occupy a considerable proportion of the cytoplasm. The cell membrane of the nerve cells is an extremely delicate structure much finer than the nuclear membrane. Electron microscopy demonstrates that it cannot be more than a few hundredths of a micron thick, whereas the nuclear membrane is about $\frac{1}{10}$ μ thick, possibly because of its nature as a double fold of the endoplasmic reticulum.

Although the cell membrane of the neuron appears under the light microscope to be very thin, under the electron microscope it appears to be made up of the same three layers with lipid in the center and protein on either side—the thickness of the membrane being 75 A. The outside of the membrane is covered by a thick layer of mucopolysaccharide. In the regions of synapses, the cell membrane is thickened and increases in electron density. In these regions in the process of transmission of stimuli, a transmitter substance such as acetylcholine is liberated in the presynaptic membrane (the one that the impulse is coming from) and this is absorbed onto a receptor protein in the postsynaptic membrane—the one that the impulse is going to—producing a potential that is conducted along the cell by altering the balance of the sodium and potassium ions. See Figs. 133–135 for illustrations of structure and action of synapses. There is some evidence from the electron microscope that the cytoplasm of nerve cells is of high water content and that the material of the axon is still more dilute. The axon itself arises from an elevated portion of the neuron called the "axon hillock" or "cone of origin" and it may give off branches or collaterals but it usually ends in arborizations known as telodendria. A short way from the cell body, the axon develops a number of sheaths and becomes a nerve fiber. If it has a well-defined myelin sheath, it is known as a myelinated nerve (for the suggested embryological origin of this sheath see Fig. 130) and if not, as an unmyelinated nerve fiber. Studies with polarized light however and with the electron microscope lead us to believe that all nerve fibers possess some type of lipoprotein sheath similar to the myelin sheath. Studies of the myelin sheath using polarized light, x-ray diffraction, and later electron microscopy (particularly in the last case the work of Fernandez-Moran should be mentioned) have shown that lipoid and protein are laid down essentially in a series of alternating concentric layers. These lipoprotein layers appear to be about 130 A thick,

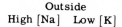

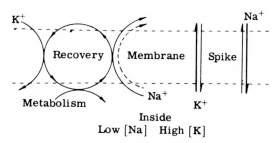

FIG. 132 Diagrammatic illustration of ion movements through the nerve membrane (after A. L. Hodgkin and R. D. Keynes, *J. Physiol.* **128,** 28, 1955). Ion movements occurring during the action wave are shown at the right; active transport, involving coupling of metabolic energy, is shown on the left. The excitable membrane pictured is presumably identical with the limiting surface membrane of the axon, or some portion thereof. (Figure and legend from Schmitt, *Rev. Mod. Phys.* **31,** 455, 1959.)

and the protein section to be about 30 Å thick. The lipid and the protein molecules of which these layers are composed are oriented radially, that is at right angles to the surface of the fiber (see Fig. 131). (See Fig. 130 for the origin of the myelin sheath.)

When hematoxylin and eosin preparations are made of nerves fixed by conventional fixatives in the usual way and embedded in wax, a reticulate structure can be seen in the myelin sheaths, which is known as the neurokeratin network. We shall have more to say about this later. At intervals the myelin sheath becomes divided at the node of Ranvier; the neurilemmal sheath is continuous at this point but the myelin sheath itself is interrupted. The reason for this is not known but it is possible that the myelin may be a type of liquid crystal; this would probably be unstable in a tube longer than that represented by the internodes between the two adjacent nodes of Ranvier. It has been mentioned that the axon is also in a semifluid condition and, under the polarizing microscope, it can be seen to be birefringent suggesting a regular molecular orientation that is not incompatible with fluidity since it represents another form of liquid crystal similar to that of the myelin sheath, but its chemical composition is different from that of the latter.

It has been found that the concentration of ions is different inside and outside the cell and the nerve fiber; sodium and chloride, for instance, have a higher concentration extracellularly and potassium intracellularly. An electrical potential exists across the cell membrane because of this difference. Both sodium and potassium diffuse across cell membranes so that presumably they would eventually become equally distributed on either side of the membrane and the membrane potential would then disappear. But this does not happen, and a factor that prevents it is a metabolic "pump," which maintains the difference in ion distribution. The level of the membrane potential depends upon the ratio of (K^+) and (Cl^-) ions and the rate at which the sodium ion (Na^+) is pumped.

The presence of a membrane potential along a nerve membrane is essential for the passage of an impulse. In the resting condition the membrane is known as the K^+ membrane. During the passage of an impulse, the potential across the nerve membrane changes, this is known as the action potential, and is due to a change in the permeability of the membrane and a sudden influx of sodium ions. The nerve fiber membrane is then called a (Na^+) membrane. The recovery of the membrane and the reestablishment of a (K^+) membrane starts with an increase in permeability to (K^+) ions and a decrease in permeability to (Na^+) ions. These changes finally bring the potential back to the resting value (see Fig. 132).

The studies on the conductance of (Na^+) and (K^+) ions in nerve were carried out by Hodgkin and Katz with the introduction of the membrane voltage clamping technique. In effect what they demonstrated was that the movement of (Na^+) into the fiber during the passage of an impulse depolarizes the membrane by carrying electrical charges through it. The reestabishment of a resting potential across the fibers (the K membrane) requires the sodium ions to be pumped out against a gradient of (Na^+). This can only be done by the use of energy. That metabolic processes are involved in this action is indicated by the relatively high oxygen uptake of nerve.

By using optical methods, it has also been demonstrated that changes in the electrical activity of the nerve fiber that are associated with the passage of a stimulus are associated with changes in the optical density or in the light-scattering capacity of the fiber. The scattering of light by the fiber indicates change of volume. It is of interest that this swelling is maintained for several seconds after the summated action of several series of impulses. The rate

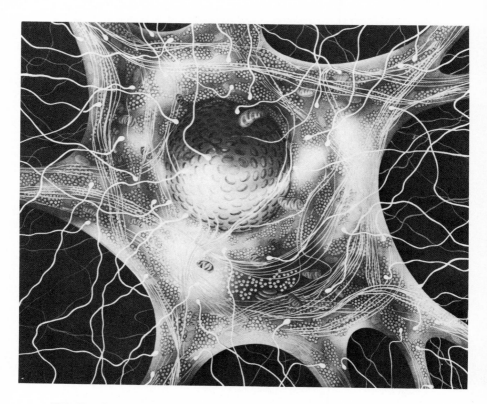

FIG. 133 Schematic structure of a nerve cell (neuron). Synaptic termina-
tions link the nerve cells. They do not adhere to the cell surface but are
separated by a 200 Å gap, which is–called the intersynaptic space. Within
the cytoplasm of the neuron can be seen the typical neurofibrils, mitochondria, pigment grounds, and tigroid bodies. (Figure and legend from
Rassegna Med. **15,** No. 3, 37, 1968.)

of change in optical density is not comparable to the velocity of
propagation of the action potential, but there appears to be a connection between the wave of excitation and the optical changes. However, the nature of this connection is obscure at the moment.

It is interesting to note that the Russian authors Kayushin
and Lyudkovskaya believed that the surface movements (as expressed
by shifts in the interference bands) result from changes in the fiber

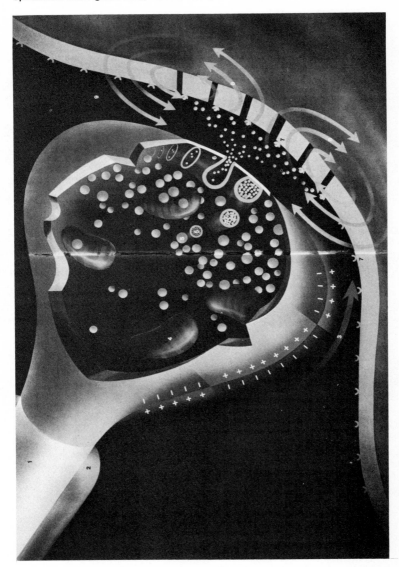

FIG. 134 Scheme of Synaptic Excitation. 1. Axon, 2. myelin sheath, 3. activating current, 4. mitochondria, 5. synaptic vesicles, 6. synaptic vesicle liberating acetylcholine, 7. post-synaptic molecular receptors, 8. post-synaptic activating currents, and 9. inter-synaptic space. The transmission of the nervous impulse is linked to a specific electrochemical process. The activating current first travels along the neuronal process, then radiates over

volume and, since these shifts disappeared when there was no stimulation, it appeared that these volume changes had returned to normal
and the movement was probably of an elastic nature. These mechanical waves (surface movements, volume changes) are presumably connected with changes of the molecular micellar structure in the nerve.
Thus, it appears that the whole of the structural system of the
nerve takes part in the conduction of an impulse. This way of looking
at the relationship between molecular structure and the excitation—conduction—recovery cycle in the nerve is an interesting approach
to the investigation of nerve physiology.

It is possible that the mechanical waves in the nerve fiber
push acetylcholine down toward the myoneural junction, and there
is some evidence this compound may be formed either in the axon
or possibly in the cell body itself. There is evidence of considerable
synthesis of protein by nerve cells, the significance of this is not
known for certain but is probably related to the production of enzyme
protein.

The fluid substance of the axon is believed to be under pressure. As evidence for this one might quote the experiments of Paul
Weiss in New York, who found that if a nerve fiber is ligated, the
axoplasm is dammed up to the point where the constriction occurs
and when pressure is released it flows forward again. He also showed
that, if a nerve fiber is cut, the axonal material will stream out
from the cut surface. It has also been shown that, if such a severed
nerve is stimulated, the speed of outflow of axoplasm is decreased.
This appears to be due to gelatinization of the axoplasm and a
consequent increase in its viscosity.

It is of interest that pressure (about 1½ lb) on the nerve
will cause an interruption of the normal function of a motor but
not of a sensory nerve, and this may last some two weeks before
coming completely back to normal. This is particularly curious when

---

the synapse and causes the liberation of the acetylcholine contained in the
synaptic vesicles. The acetylcholine traverses the intersynaptic space and
becomes attached to molecular receptors situated on the post synaptic
membrane (the cell membrane of the neuron to be activated) and this
causes a postsynaptic current of action in this membrane. The free acetylcholine liberated is rapidly destroyed by the enzyme acetyl cholinesterase.
The energy necessary for these processes is furnished by the mitochondria.
(Figure and legend from *Rassegna Med.* **15**, No.3, 38, 1968.)

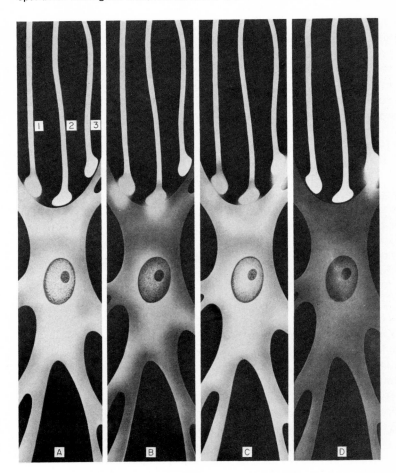

FIG. 135 Both excitatory and inhibitory types of synaptic action exist. The diagram shows two excitatory synapses (1, activated) and (2, not activated), and an inhibitory synapse (3, not activated). In the case of a motor nerve cell a single excitatory impulse is not sufficient to cause any motor response from the cell (A). The response occurs only when both excitatory synapses enter into action (B). Should the inhibitory synapse also enter into action the threshold is not crossed so there is not a motor response (C). In (D) the inhibitory synapse alone is in action. From an electrical point of view, there is no difference between the excitatory or inhibitory types of impulse which travel along nerve fibers. Inhibition or excitation depend exclusively on which kind of chemical agent is released by the electrical impulse. The excitatory synapse liberates acetylcholine. Gamma amino butyric acid is

such pressure produces no microscopically visible degenerative changes in the nerve fiber.

The cell membrane in nerve fibers is complicated because of the close association of the axon with the Schwann cell. The Schwann cell is associated with the axon in both myelinated and nonmyelinated fibers. In the case of the nonmyelinated fibers, the axon is embedded to varying degrees into the protoplasm of the Schwann cell, but is still surrounded closely by the cell membrane of the Schwann cell that it has invaginated with it. The actual distance between the outer parts of the two membranes under these circumstances is approximately 150 Å. The membrane of the axon and the membrane of the Schwann cell are each approximately 75 Å and are made up of three parts. There are two rather dense areas that are about 25 Å across, and these are separated from each other by a less dense area also about 25 Å across. So that we may consider that the membrane of the nerve fiber is enlarged to this extent by the membrane of the Schwann cell itself.

In myelinated fibers the axon instead of just being embedded in the cytoplasm of the Schwann cell is surrounded by many layers of Schwann cell cytoplasm and cell membrane. The mechanism that produces this has been demonstrated by Geren in tissue cultures of developing nerve fibers. She demonstrated that the Schwann cell winds in a spiral fashion around the axon thus building up many layers of Schwann cell membrane around the axon itself. Many degrees of myelination of nerve fibers are known, and the degree of myelination depends on the number of times the Schwann cell winds around the axon (see Fig. 130).

Electron-microscope studies of myelinated nerve fibers have demonstrated very complex arrangements of the myelin so produced. The first of the high-resolution studies of such fibers were produced by Fernandez-Moran and Sjöstrand. If a myelinated nerve is first fixed in osmium tetroxide (osmic acid), it could be seen that the myelin was made up of a series of lines which absorbed the osmium very strongly. These lines were approximately 25 Å thick and repeated at a period of up to 120 Å. Each of these dark lines was separated by a light line of approximately the same size, and running

---

believed to be the chemical agent with an inhibitory action. (Figure and legend from *Rassegna Med.* **46,** No. 1, 40, 1969).

across the center of each light line was a line that was more dense than the light line but less dense than the main 25-Å repeating lines. In interpreting the significance of these lines, Robertson has pointed out that the gap, which in the unmyelinated fibers was between the two cell membranes and measured 75 Å across, is largely eliminated in myelinated fibers and the outer part of the membrane of the Schwann cell comes into direct contact with the membrane of the axon. Thus we can say that, as Robertson points out, each of the repeating units that are obvious in electron micrographs of the myelin sheath is two Schwann cell membranes in contact along their outside surfaces. The Schwann cell membrane probably consists of a single bimolecular leaflet of lipids of which the hydrophilic ends are associated with monolayers of a nonlipid material and which is probably protein, although there is a possibility that either some polysaccharide or glycoprotein of some sort may be associated with it, possibly on the side directed toward the cytoplasm.

We have mentioned that there is a high concentration of potassium ions inside the nerve fiber in the axon and much less concentration of sodium. Now it appears that sodium continuously diffuses into the nerve fiber and it must be excreted from the axon to the exterior again. The difficulty about this is that the sodium has to be excreted against a gradient, that is to say, sodium has to be passed out of the axon into a medium where there are more sodium ions than there were in the axon and vice versa with potassium (see Fig. 132). This process requires energy and there have been two principal views as to how this could be done. The two theories are the redox pump theory and the sodium pump theory.

Conway has suggested the following way in which the outward movement of sodium ions in the nerve fiber could be explained. Sodium ions could be continuously passed on from unsaturated phospholipid molecules to the next unsaturated phospholipid through the membrane to the outside, and there would be alternate oxidation and reduction of the phospholipid molecules as this happened. The other view is that sodium is pumped out by energy produced by presence of ATP, creatine phosphate, and oxidative enzymes and possibly also with the aid of phospholipid. What is not certain yet is whether Na and K ions need to be passed through the myelin sheath to maintain the physiological functioning of the nerve fiber. This is a problem for which we will no doubt get the answer in due course.

It is difficult to draw many conclusions about the division of labor in the nerve fiber. First of all, in the nerve cell we have great production of protein and possibly the production of acetylcholine and the passage of one or both these compounds down the axon toward the myoneural junction; the passage of the nerve impulse is provided for by the axon membrane itself. However, the precise relationship of the myelin sheath to the propagation of the nerve impulse is not known. In myelinated nerves, the impulse appears to be faster than in nonmyelinated or poorly myelinated nerves, and yet one does not know quite in what way the myelin sheath is able to do this. At the termination of the nerve fiber, there is a modification in the form of complex branching in association with the sarcolemma to build up the myoneural junction. There is an accumulation here of acetylcholine which presumably passes through the membrane of the neural part of the mechanism to stimulate the formation of a contraction wave by its influence on the sarcolemmal membrane. The presence of cholinesterase in these junctions, which is necessary for inactivation of acetylcholine after it has done its job, is of interest but its precise origin is unknown. So we can see evidence of division of labor in the nerve cell and its axon, but the nature of the labor performed by each part is still partly speculative.

EIGHT
Conclusion

WE have studied the structure and to some extent the chemical composition and the biochemical composition of the cell in general and of three specialized types of cells. We have demonstrated that various parts of the cell perform very important but different functions, that, in other words, there is a division of labor. Each part performs the job for which it is chemically and structurally suited; thus the nucleolus synthesizes protein and ribonucleic acid. A large part of the protein synthesis of the cell is carried out by the ribosomes associated with the endoplasmic reticulum of the cytoplasm. We know that the chromosomes of the nucleus contain the genes that produce hereditary effects in the cell. We have seen that the oxidation activities of the cell are carried on in the mitochondria and that, as a result of respiratory activity of these structures, ATP is produced, which is available for energy and structural purposes in the cell. In the glandular cell, the ATP provides energy for the synthesis of the products of secretion. In the muscle fiber, it provides the energy for muscular contraction; and presumably in the nerve fiber the myelin sheath provides the energy for the extrusion of Na ions through the nerve membrane. The cytoplasmic (endoplasmic) reticulum in addition to synthesizing protein may have other activities. If, as is believed by some authors, the cisternae in the endoplasmic reticulum are continuous with the outside we have, in fact, an enormous increase of area of the membrane of the cell. We have the outside world actually penetrating deep into the cytoplasm and even surrounding the nucleus so that some of the nuclear products could be secreted directly to the exterior of the cell without actually passing through the cytoplasm itself. Conversely, glucose could pass directly into the nucleus as well as directly into the interior of the cytoplasm.

What is so striking in the intact cell is not that there is a division of labor but, what is more fantastic, the degree to which all the various parts of the cell cooperate with each other to build up a metabolic picture that is ordered and controlled. This cooperation rarely gets out of gear, and when it does disorders of growth and metabolism occur, of which one variety can be cancer. The mechanism of cell function is under control of the endocrine system, and we have mentioned earlier in this book how the fine structure of the prostate cell is under control of the male sex hormone and how the localization of acid phosphatase in the same cells is also under control of the same hormone. This is a field that is still scarcely touched and is likely to produce, in the future, an enormous amount of interesting and valuable information about the function of the cell.

The relationship of vitamins and nutritive factors in general to cell structure is also only beginning to be investigated. Studies on starvation have demonstrated a great increase in the number of mitochondria, and our own studies on scurvy have demonstrated the more intimate association of the endoplasmic reticulum with the mitochondria with a simultaneous increase in the number of mitochondria. Here is a vast field in nutrition where electron microscopy would probably produce immense contributions to a study not only of the physiology of the cell but of the function of vitamins and other nutrients in cell metabolism.

The precarious nature of the cell structure is demonstrated by the fact that the structural integrity of mitochondria is affected in the absence of ATP, and it seems possible that this whole complicated system of energy production is essential not only to provide for cell activity but also energy to maintain the structural integrity of the cell and its organelles; if this is so, then any serious interference with the ATP production cycle will lead to a breakdown of the complex cellular structure as well as interference with the metabolism.

New techniques of cytological investigation have in the last few years made fundamental alterations in our outlook on cell structure and function, and probably the most fascinating development is our knowledge of DNA and RNA. The process of breaking the code of DNA is providing the kind of information and technique for actually altering the genetic material. This intriguing possibility has been engaging the minds of a number of leading workers in this field. Characteristics of bacteria and fungi have been altered by bringing them into contact with other DNAs. The synthesis of artificial DNAs and RNAs was performed in experiments to crack the code for the alignment of amino acids, and this opens the way for controlling heredity or even producing entirely new forms of life. One of the problems still to be solved is the differential action of DNA. The original fertilized egg simply has a supply of DNA. As it multiplies the same DNA simply becomes replicated in the various cells. Up to a certain point any of these cells if separated from the others has the potential to produce a whole embryo. Identical twins, for example, are produced by the separation of the two cells formed by first division of the fertilized egg.

Beyond a certain point the cells of a dividing embryo lose the ability to reproduce the whole embryo—this is because some of the genes have been "switched off." Similarly the cells formed from the first division of the fertilized eggs all look alike and later they change their form and character, becoming kidney cells, muscle cells, and so on. Again this must be due to the "switching on" or "off" of certain genes in the DNA. How is this switching done? This is the most intriguing development in this area that awaits solution and it is one in which the next major breakthrough will probably be made. Already an enzyme called "replicase" has been found, which may give a hint of the mechanism. This enzyme is released in a cell when it is invaded by viruses; it has the effect of switching off the cell's own genetic mechanism and leaves it to follow the virus's genetic instructions, which are to make new viruses.

As man gains control of his genetic material and the switching mechanism, many exciting but also worrisome possibilities open up. For example, a piece of an organ of some person with specially desirable qualities could be cultured by standard tissue-culture procedures. In such cells all the genes except those for making that particular organ would have to be switched off; if they could be switched on again then each of the cells in that tissue culture would have the ability to reproduce the entire person from which the original tissue was obtained.

Further knowledge of the functioning of DNA will also shed light on the mechanism of aging and perhaps suggest methods of ameliorating its effects. For example, it has been suggested that one of the mechanisms of old age is the development of a defect or defects in the DNA. This may have happened because of the accumulation of enzymes like replicase in the cells that partly obscure the genetic mechanism or because of damage to the DNA by cosmic radiation or for various other reasons known or unknown. Perhaps old persons will be returnable to some more youthful condition by treatment with an undamaged DNA or the cell cleared of materials that hinder its genetic mechanism.

While all these ideas sound interesting and fascinating, most of them depend upon the belief that all the hereditary material of the cell is in the nucleus. But even in heredity there is division of labor. The nucleus plays its part, but in the cytoplasm of the cell there is the cytoplasmic inheritance of the mitochondrial DNA. All the imaginings and dreams of controlling man's form and destiny have to take both systems into account and the mechanism of such control is probably much more complex than we have yet conceived.

Subject Index

A

Acetabularia mediterranea, 225,
228–229, 231
life cycle, 222–223
nucleus and, 223
Acetylcholine, nerve fiber and, 279
Acetyl coenzyme A, 113, 119
fatty acids and, 130, 131
Acid phosphatase, see also Phosphatase(s), lysosomes and,
103, 105
Aconitase, 113
Actin, 260
composition of, 262, 264
localization of, 262
muscle contraction and, 264
Adenosine diphosphate, respiratory
rate and, 127–128
Adenosine 3′,5′-monophosphate,
hormones and, 211
Adenosine triphosphatase, cilia
and, 32
meromyosins and, 262
permeability and, 27, 31
Adenosine triphosphate, chloroplast and, 100–101, 103
muscle contraction and, 264,
266, 268–269
nucleus and, 164–165
production of, 91–92, 115–117
respiration and, 127–128
utilization, nucleus and, 227–
228
Adrenal cortex, mitochondria of, 77
Alanine, formation of, 112, 137
Aldolase, 112, 115, 117, 163
Alkaline glycerophosphatase, see
also Phosphatase(s) Golgi apparatus and, 147
Alkaline phosphatase, Golgi apparatus and, 152, 154
Amblystoma, liver, mitochondria
of, 81

Ameba, anucleate, 221–222, 224,
226–228, 229
artificial, formation of, 11
cell membrane of, 19
hybrid, antibodies and, 231
nuclei, transplantation of, 220–
222, 229–230
Amino acid(s), cell permeability,
24, 25
degradation of, 133–135
genetic code and, 208–209
primitive conditions and, 2
transfer ribonucleic acid and,
207
Ammonia, primitive atmosphere
and, 2–4
Ammonium thiocyanate, primitive
conditions and, 4
Amoeba proteus, pinocytosis in,
29
Arthritis, lysosomes and, 106
Atmosphere, primitive, 2

B

Bacteria, origin of mitochondria
and, 66, 209
Bacteriophage, infection by, 194,
202, 203, 204

C

Calcium ions, muscle contraction
and, 256–257, 266
Calliphora, flight muscle of, 259
Calvin cycle, carbon dioxide fixation and, 101
Cancer cell, Golgi apparatus of,
150
Carbohydrate(s), metabolism of,
108–129
Carbon dioxide, fixation in photosynthesis, 101
Carotene, penetration of cell by, 23

Carotenoids, quantasomes and, 99–100
Castration, endoplasmic reticulum and, 121
  Golgi apparatus and, 154
Caveolae intracellulares, 30
  endoplasmic reticulum and, 58, 61–62
  muscle and, 248
Cell(s), damage, lysosomes and, 106
  division of labor in, 234, 284–287
  structure of, 7–10, 59
Cell membrane, see also Membranes
  electron microscopy, 16–19
  endoplasmic reticulum membranes and, 58–64
  permeability, compounds, 22–25
    ions, 25–28
    pinocytosis and, 28–34
  structure of, 12–21
Centriole, structure of, 177–179, 181
Chick embryo, cells of, 8
Chloramphenicol, protein synthesis and, 217
Chlorophyll, origin of life and, 5
  quantasomes and, 99
Chloroplasts, nucleic acid in, 209
  size and shape, 97–99
  structure of, 98–101, 102
Cholesterol, membrane formation and, 6, 20
  mitochondria and, 81
Cholinesterase, motor end plates and, 249, 250
Chromatin, ergastroplasm and, 46, 47
Chromosomes, lampbrush, ribonucleic acid and, 173, 232
  structure of, 172–177
Cilia, structure of, 31–34
Cisternae, endoplasmic reticulum and, 58–60
Coenzyme Q, 115
  mitochondria and, 91–93

Collagen, reaggregation of, 7
Condensine enzyme, 113
Cortisone, lysosomes and, 106
Creatine phosphatase, permeability and, 27
Cristae mitochondriales, cytochrome and, 71
  surface area of, 72–73
Cyclops, muscle of, 265
Cytochrome(s), oxidations and, 114–115
  synthesis of, 83, 84
Cytochrome c, mitochondria and, 91
  nucleus and, 163
Cytochrome oxidase, 114, 115
  mitochondria and, 82, 84, 93
  permeability and, 27
Cytochrome reductase, 114–115, 117, 120
Cytopemphis, 30
Cytoplasm, composition of, 36–37
  relationship to nucleus, 219–235
Cytoskeleton, 40, 42

D

Deoxyribonucleic acid, composition, species and, 191
  distribution of, 182
  heredity and, 191–194, 204
  messenger, 218–219
  mitochondria and, 73, 83–84, 209–211
  molecular structure, 183, 188–190, 206
  nucleolus and, 168, 170, 171
  transcription, mechanism of, 197–198
Detoxification mechanisms, 25
Differential centrifugation, 34–35
Dinitrophenol, respiration and, 127
Dotterkern, 46
Duodenal cell, Golgi apparatus of, 145

E

Ectoplasm, 40

Egg, newt, mitochondria of, 92
Electron transfer, 92–94, 115
   chloroplast and, 104
   phosphorylation and, 89, 91
Endoplasmic reticulum, carbohy-
      drate metabolism and, 118,
      120–121, 124, 127
   membranes, cell membrane and,
      58–64
   nucleus and, 233
   structure of, 43–55
Enolase, 112
Enzymes, Golgi apparatus and,
      152–160
   lysosomes and, 103, 105
   mitochondrial, 84, 87
   muscle and, 254, 268–269
   nucleolus and, 212–213
   nucleus and, 162–164, 219–220
   peroxisomes and, 106
   synthesis, nucleus and, 230–
      231
Epididymal cell, Golgi apparatus of,
      144, 150
Epithelial cell, Golgi apparatus of,
      142
Ergastoplasm, 43
   endoplasmic reticulum and, 44
   nature of, 45–51
Escherichia coli, protein synthesis
      by, 217
Euglena, chloroplasts of, 99

F

Fat, metabolism, 129–133
   nucleus and, 163
   muscle and, 252–253
   synthesis of, 131
Fatty acid(s), cell permeability, 24
   oxidation, sarcosomes and, 269
   synthesis of, 131, 132–133
   unsaturated, uncoupling by, 128
Fatty acid oxidase, mitochondria
      and, 84
Ferredoxin, photosynthesis and,
      101, 103

Ferritin, storage by mitochondria,
      75
Fibrils, muscle, 258–260
   composition of, 260–266
Flagellum, structure of, 31–34
Flavin adenine dinucleotide, mito-
      chondria and, 83
Flavoproteins, oxidations and, 114
Frog, sartorius muscle of, 251, 261
Fructokinase, 111
Fructose 6-phosphatase, 163
Fumarase, 113

G

Genes, nucleic acids and, 182–191
Genetic code, nature of, 205–209
Gliadin, membrane formation and,
      6
Gluconic acid, formation of, 110
Glucose, cell permeability, 24, 27–
      28
   oxidation, 88, 109–115
      control of, 120–121
      muscle and, 268–269
Glucose 6-phosphatase, 110–111,
      117, 118, 120
   localization of, 122–124
Glucose 6-phosphate dehydrogen-
      ase, 110, 117
Glycerol, cell penetration by, 23,
      24
α-Glycerophosphate, lipids and,
      112
Glycogen, formation of, 109, 111,
      112, 120, 135
   sarcoplasm and, 250, 252
   utilization, nucleus and, 227,
      228
Glycogen synthetase, activation of,
      125
Glycolysis, 90, 109–112
   localization of, 117
   nucleus and, 163, 219
Glyoxylate cycle, peroxisomes and,
      106
Goblet cells, Golgi apparatus of,
      159

Golgi apparatus, endoplasmic reticulum and, 63
enzymes in, 152–160
function of, 141, 146, 147, 151
muscle, 258
nerve cell, 272
secretions and, 154–155, 159,
236, 240, 241, 244–246
structure of, 140–146, 147, 151
Grana, chloroplasts and, 98, 99
Gulonolactone oxidase, primates
and, 208

H

Helix, Golgi apparatus of, 148
Hemoglobin(s), amino acid se-
quences of, 199–202, 208
Hexoestrol phosphatase, 148
Hexokinase, 109–110, 117, 118,
120, 121, 125
Hormones, cell enzymes and, 210–
212
Hyaloplasm, 37, 41–42
Hydrogen cyanide, primitive at-
mosphere and, 2, 4
Hydrogen sulfide, primitive atmos-
phere and, 2, 4

I

Insulin, permeability and, 28
Ions, penetration of membranes,
25–28
Isocitric dehydrogenase, 113

J

Janus green B, mitochondria and,
78–79

K

α-Ketoglutaric dehydrogenase, 113
Kidney cells, Golgi apparatus of,
150, 156
mitochondria of, 93

pinocytosis by, 30
ribonucleic acid granules of, 186
Kinetochore, chromosomes and,
175
Krebs cycle, 113–114, 135
mitochondria and, 91, 119, 130

L

Lactic acid, accumulation, work
and, 108–109, 269
cell permeability to, 24
Leucocyte, mitochondria of, 94,
95, 96
Lipid, cell membrane and, 12–16,
19–20
Golgi apparatus and, 148–149,
155, 156
mitochondria and, 78, 79–80,
81
Lipofuscin, 272, 274
Lipoid, definition of, 12
Littorina, plasma membranes, 18
Liver, endoplasmic reticulum of,
49, 50
Golgi apparatus of, 158
microsomes, 56
plasma membrane, 17
Lymnaea, mitochondria of, 79
Lymphocyte, endoplasmic reticulum, 53
Lysosomes, composition of, 103,
105
endocytosis and, 105, 107

M

Macrophage, endoplasmic reticulum, 43–44
Maize, chloroplast of, 98
Malic dehydrogenase, 113
Melanin, nerve cells and, 272
Membranes, see also Cell membrane
formation, origin of life and,
6–7, 9, 11
mitochondrial, 91–92, 95–97

nerve cell, 274, 281
  ion movements through, 275–276
  nuclear, 165–166
Membrane potential, nerve impulse and, 276
Meromyosin(s), arrangement of, 262
Meteorites, purines and pyrimidines in, 1
Methane, primitive atmosphere and, 2–4
Methyl alcohol, penetration of cell by, 23
Microsomes, 38–39
  enzymatic activity of, 237–238, 240
  nature of, 55–58
Microtubules, structure of, 180
Mitochondria, 39
  abnormal storage by, 75–76
  chemical nature, 78
  composition of, 74
  control of metabolic rate and, 119–120
  endoplasmic reticulum and, 63–64
  ergastoplasm and, 47
  fatty acid oxidation by, 130
  metabolic substances in, 78–97
  movements of, 67, 226
  nerve cell, 272
  nucleic acids in, 73, 83–84, 209–211, 217
  origin of, 66, 73
  permeability of, 74
  reconstruction of, 85
  secretions and, 245–246
  size and occurrence of, 65–67
  structure of, 70–73
  synthetic activity, 76–77
Mitochrome, uncoupling and, 128–129
Motor end plates, 250
  sarcolemma and, 249
Mucopolysaccharides, cell membrane and, 19

Muscle, see also Striated muscle fiber
  endoplasmic reticulum of, 50
  mitochondrion of, 70
Myelin, formation of, 273, 274–275
Myelin sheath, structure of, 14, 21–22, 281
Myoglobin
  distribution of, 253
  sarcosomes and, 268
Myosin, 260
  localization of, 262

N

Nebenkern, 46
  endoplasmic reticulum and, 63
  structure of, 50–51
Nerve cell, Golgi apparatus of, 272
  mitochondria, 272
  structure of, 270–275, 277, 281–282
Nerve conduction, water and, 42
Neurofibrils, nature of, 270–272
Nicotinamide adenine dinucleotide, dehydrogenases and, 110, 112
  mitochondria and, 83, 92–93
  reduction, respiratory rate and, 126–127
  synthesis, nucleus and, 163, 228, 233
Nicotinamide adenine dinucleotide phosphate, photosynthesis and, 101, 103
Nissl bodies, 46, 51, 270
  composition of, 272
Nucleic acid(s), bases, primitive conditions and, 2, 5
  structure of, 183–191
Nucleolonema, structure of, 170
Nucleolus, mitochondria and, 67–68
  protein synthesis by, 212–219
  ribonucleic acid, particles, 51, 55
  synthesis, 231
  structure of, 168, 170–171, 180

5'-Nucleotidase, Golgi apparatus and, 154–155
Nucleotides, metabolism, nuclear enzymes and, 163, 220
Nucleus, cytoplasmic relationships, 219–235
  endoplasmic reticulum and, 55, 61, 63, 121, 124
  glucose and, 124
  muscle fiber, 258
  nucleic acids of, 191–212
  protein synthesis in, 232–233
  ribosomes and, 55
  size, shape and position, 161–162
  structure of, 164–169

O

Oestrone phosphatase, Golgi apparatus and, 149
Opalina, cilia of, 32
  mitochondria of, 76
Oxalosuccinic dehydrogenase, 113
Oxidases, mitochondria and, 79, 81
Oxidation, rate, control of, 125–127
β-Oxidation, fatty acids and, 130
Oxidative cycle, nucleus and, 163, 219
Oxygen, penetration of cell by, 23, 24

P

Pancreas, secretion by, 237–241
Pancreatic cell, endoplasmic reticulum, 54, 60
  Golgi apparatus of, 143, 154
  secretion droplets in, 243
Paramoecium, mitochondria of, 75
Parotid gland, Golgi apparatus of, 153
Periodic acid Schiff reaction, muscle and, 250, 252
Permease(s), induction of, 28
Peroxisome, 55
  composition of, 106

Phenylketonuria, 208
Phosphatase(s), see also specific enzymes
  Golgi apparatus and, 155–160
  nucleoli and, 212–214
  permeability and, 26, 27–28, 31
  salivary gland, 242–243
Phosphate, uptake, nuclear control of, 227, 229
Phosphatidic acid, permeability and, 27
Phosphofructokinase, 111–112
Phosphoglucomutase, 111, 117, 121
Phospholipids, mitochondria and, 81
  quantasomes and, 99–100
  synthesis of, 131
Phosphorylase, 117, 121
Phosphorylase B-kinase, 121
Phosphorylation, oxidative cycle and, 115–117
  primitive conditions and, 5
Pinosytosis, 24, 29, 30
  caveolae intracellulares and, 62
Plant cells, metabolism in, 137, 139
Plasma cell, endoplasmic reticulum, 52
Plastids, function of, 97
Podocytosis, 30
Polypeptides, primitive conditions and, 4
Polysaccharide, sarcolemma and, 249
Polysomes, 55
Pongo pygmaeus, chromosomes of, 175
Pores, nuclear membrane, 165–166, 202–203
  plasma membrane, 21, 28
Potassium ions, nerve fiber and, 276, 282
Primeval conditions, organic compounds and, 1–4
Protease, salivary gland, 242
  trypsin activatable, microsomes and, 237–238, 240

Protein(s), cell membrane and, 13–16, 19–20
  contractile, permeability and, 27
  extracellular, 215
  gel formation by, 41–42
  Golgi apparatus and, 149–150
  metabolism, 133–137
    nucleus and, 163
  mitochondrial, 81
    synthesis of, 83–84, 230
  nucleic acids and, 191
  quantasomes and, 99–100
  structure of, 198–199
  synthesis, mitochondria and, 83–84, 230
    nerve cells and, 272
    nucleic acids and, 192, 193, 195–197, 207
    nucleus and, 230–233
    nucleolus and, 212–219, 231
Protoplasm, alveolar theory, 38
  filar and reticular theories, 38
  granule theory, 39
Pseudotrichonympha, cilium of, 33
Pyrenoids, 99
Pyruvate, oxidation of, 109, 113–115, 118–119
Pyruvate oxidase, 163

Q

Quantasomes, composition of, 99–100

R

Red cell, cholesterol of, 21
  ghost, 13
  membranes, 13–14, 16, 20
Respiration, nucleus and, 224–225
  salivary gland, 243–244
Respiratory rate, tissues and, 115
Rhopheocytosis, 30
Ribonuclease, cell permeability and, 28
  microsomes and, 237–238, 240
  penetration of cell by, 220
Ribonucleic acid, chromosomes and, 173

distribution of, 182
  messenger, 197
    synthesis of, 197–198, 206–207, 218, 219
  mitochondria and, 83–84, 210
  nucleolus and, 171, 232
  production, nucleus and, 224
  transfer, 197, 207, 216
Ribonucleoprotein, endoplasmic reticulum and, 44, 50, 51, 57
Ribose, primitive conditions and, 5
Ribosomes, endoplasmic reticulum and, 48–49, 51, 55
  formation of, 218, 219
  protein synthesis and, 216

S

Salivary gland, chromosome of, 178
  secretion by, 241–244
Sarcolemma, structure of, 247–249
Sarcoma cells, mitosis of, 10
Sarcoplasm, composition of, 249–255, 258
Sarcoplasmic reticulum, 256, 257
  structure of, 255
Sarcosomes, structure of, 266–268
Schwann cell, nerve fibers and, 281–282
Scurvy, mitochondria and, 217
Secretion, 239
  endoplasmic reticulum and, 60–61, 237–241
  Golgi apparatus and, 154–155, 159, 236, 240, 241, 244–246
Semnopithecus, ovarian cells, Golgi apparatus of, 147
Senescence, lysosomes and, 106
Sickle-cell anemia, hemoglobin and, 200, 201–202, 208
Sodium ions, nerve fiber and, 276, 282
Spermatid, Golgi apparatus of, 150
Spinach, chloroplasts, 100
Spinal ganglion cell, Golgi apparatus of, 141

Spinal ganglion cell—Continued
    mitochondria of, 67–68
    nucleolus of, 176–177
Spindle fibers, structure of, 179
Spirogyra, chloroplasts of, 99
    nucleus, removal of, 223
Starfish, eggs, surface tension of, 13
Starvation, endoplasmic reticulum and, 237
Stentor, enucleate, 225
Striated muscle fiber, see *also*
    Muscle contraction of, 256–258, 264–266, 267
    enzymes of, 254, 268–269
    structure of, 246–247
Succinic dehydrogenase, 113
    mitochondria and, 82, 84
    salivary gland, 243
Sulfhydryl groups, mitochondria and, 80–81
Synapse, nature of, 274
Synaptic excitation, 276, 280
Synaptic inhibition, 280

T

Tenebrio, flight muscle, 263
Thiamine pyrophosphatase, Golgi apparatus and, 159–160
Tobacco mosaic virus, heredity in, 212
Transaminases, mitochondria and, 83, 135
Transformation, deoxyribonucleic acid and, 194

Trigeminal ganglion cells, Golgi apparatus of, 157
Tropomyosin, 264
    localization of, 262
    muscle contraction and, 266
Troponin, muscle contraction and, 266

V

Virus, protein coat of, 201
Viscosity, protoplasm, 37
Vitamin A, mitochondria and, 79–80
Vitamin B_{12}, protein synthesis and, 215–216
Vitamin(s) B, mitochondria and, 80
Vitamin C, biosynthesis of, 207–208
    Golgi apparatus and, 150–152, 156
    mitochondria and, 80

W

Water, intracellular, 42

X

Xenopus laevis, nucleoli of, 168

Z

Zymogen granules, starvation and, 237, 238